WJTA Conference and Expo 2023

AA006148

New Orleans, Louisiana, USA
30 October - 1 November 2023

ISBN: 978-1-7138-9779-8

Printed from e-media with permission by:

Curran Associates, Inc.
57 Morehouse Lane
Red Hook, NY 12571

Some format issues inherent in the e-media version may also appear in this print version.

Printed with permission by Curran Associates, Inc. (2024)

For permission requests, please contact WaterJet Technology Association-Industrial & Municipal Cleaning Association at the address below.

WaterJet Technology Association-Industrial & Municipal Cleaning Association
906 Olive Street – Suite 1200
St. Louis, MO 63101-1448
USA

Phone: (314) 241-1445
Fax: (314) 241-1449

www.wjta.org

Additional copies of this publication are available from:

Curran Associates, Inc.
57 Morehouse Lane
Red Hook, NY 12571 USA
Phone: 845-758-0400
Fax: 845-758-2633
Email: curran@proceedings.com
Web: www.proceedings.com

TABLE OF CONTENTS

2023 WJTA Conference and Expo

October 30-November 1, 2023 • New Orleans, Louisiana

Paper

CHOOSING BRAKES FOR HIGH-PRESSURE ROTARY TOOLS: CENTRIFUGAL BRAKE VS. MAGNETIC SPEED CONTROL

Comet do Brasil Industria e Comercio de Equipamentos Ltda (Lemasa)
Indaiatuba, Sao Paulo, Brazil

INTRODUCTION

In the world of high-pressure water jetting, where precision, power, and adaptability are paramount, the selection of rotary tools equipped with the most suitable braking mechanism becomes a crucial decision. Two dominant solutions have emerged, each with its distinct advantages and limitations. One approach employs a Centrifugal Brake driven by centrifugal forces, while the other harnesses the power of magnets through a magnetic brake system. The choice between these systems profoundly influences the efficiency, effectiveness, and versatility of rotary tools in a range of applications across various industries.

Industries such as construction, manufacturing, and maintenance continue to demand powerful tools that excel in challenging conditions, making the decision between hydraulic and magnetic brakes even more pertinent. From the intensive surface preparation required in shipyards and dry docks to the maintenance of structures within refineries, the proper treatment and rejuvenation of metallic surfaces are critical. In such scenarios, the rotary tools' braking mechanisms play an essential role in ensuring precision, control, and safety.

SURFACE PREPARATION IN CHALLENGING ENVIRONMENTS

Consider the demanding environment of shipyards and dry docks, where metal structures face exposure to the corrosive effects of seawater and harsh weather conditions. The need for efficient surface preparation, renewal, and protective coating becomes paramount to maintain structural integrity. The choice of rotary tool brake can significantly impact the quality and longevity of the treatment. The Centrifugal Brake's stability and reliability make it well-suited for such environments, providing consistent operation even in adverse conditions.

Similarly, refineries and other industrial structures also require regular maintenance to prevent corrosion and deterioration. The intricate network of metal components demands careful surface preparation before recoating or maintenance work. Here, the magnetic brake's rotation control becomes crucial, enabling operators to achieve precise and uniform surface treatment. However, factors like the presence of metallic particles in the environment and potential temperature variations need to be weighed against the magnetic brake's effectiveness.

ROLE OF ROTATION CONTROL AND APPLICATION-SPECIFIC NEEDS

Rotation control is particularly essential in scenarios where uniform surface preparation is vital. The ability to adjust the tool's rotation enables operators to customize the intensity of water jet impact, catering to different surface conditions and treatment requirements. This adaptability finds resonance in the repair and renewal of structures within shipyards, docks, and refineries, where varying degrees of wear and corrosion necessitate different treatment intensities.

Furthermore, the application of high-pressure rotary tools extends beyond surface preparation. From the removal of coatings and corrosion to the precision cutting of metallic components, the versatility of these tools is a hallmark of their utility. In such cases, the choice between hydraulic and magnetic brakes influences not only the quality of surface preparation but also the efficacy of the overall task.

CENTRIFUGAL BRAKE: ADVANTAGES AND DISADVANTAGES WITH SPECIFIC APPLICATIONS

The rotary tool equipped with a Centrifugal Brake addresses concerns related to eddy currents and magnet heating. However, it's important to note that Centrifugal Brake systems require a larger quantity of oil, which needs periodic replacement—typically between 20 and 50 hours, depending on rotation and operational conditions. This increased oil usage not only escalates costs but also contributes to a greater environmental impact.

APPLICATION OF EPOXY PAINT REMOVAL AS AN EXAMPLE OF THIS MODEL'S USE

An exemplary application for the Centrifugal Brake model is epoxy paint removal from surfaces. Epoxy paint is known for its durability, making its removal challenging. In this context, the tool's maximum rotation is often required to effectively wear down the epoxy paint. The Centrifugal Brake, although lacking fine rotation adjustments, ensures the tool consistently operates at higher rotations, facilitating efficient epoxy paint removal.

Furthermore, epoxy paint removal may generate loose metallic residues on the surface. These residues can hinder the proper functioning of the magnetic brake model by interfering with magnetic components and affecting their efficiency. The Centrifugal Brake, not reliant on magnetism, circumvents this issue and maintains functionality in environments with potential loose metallic particles.

MAGNETIC BRAKE: ADVANTAGES AND DISADVANTAGES WITH CONSIDERATIONS ABOUT EDDY CURRENTS

The rotary tool with a magnetic brake offers the ability to adjust rotation by altering the proximity of a copper ring to the magnets. However, eddy currents are an electromagnetic phenomenon that generates heat when a conductor, such as the copper ring, is exposed to a varying magnetic field, like the field created by the magnets. This can result in heating of the copper ring and the magnets, leading to decreased effectiveness of the magnetic brake as temperature increases. At high rotations or in hot environments, this heating can become significant, impairing the brake's ability to control tool rotation.

CONCLUSION

In the realm of high-pressure water jetting and surface preparation, the choice between hydraulic and magnetic brakes holds the potential to significantly impact operational efficiency, safety, and environmental considerations. As industries continue to grapple with the challenges of maintaining metallic structures amidst harsh conditions, the suitability of the chosen brake mechanism takes on paramount importance. The complex interplay of factors, including rotation control, environmental conditions, maintenance requirements, and cost-effectiveness, guides decision-makers in selecting the optimal solution.

Whether in shipyards, dry docks, refineries, or other environments requiring meticulous surface preparation and structural renewal, the proper selection of rotary tool brakes is essential. As we delve into the advantages and limitations of each braking system, this article endeavors to empower industry professionals to make informed decisions that align with their specific needs. By fostering a comprehensive understanding of hydraulic and magnetic brakes' attributes, we aim to equip professionals with the insights needed to navigate the challenges and opportunities inherent in surface preparation across a multitude of industries.

HISTORICAL ADVANCES IN WATERBLASTING PUMPS

Frank Mullins
Jetstream of Houston, LP
Houston, Texas, USA

The focus of the present paper will be on reciprocating power pumps, specifically plunger pumps used in waterblasting. While acknowledgement may be given to the progress in developing UHP Intensifier pumps, as well as high volume Hydraulic fracturing pumps, the concentration here is specific to the development of pumps used in industrial cleaning and surface preparation applications.

WJTA has made recommendations on color coding to visually identify pressure plateaus as part of their *Industry Best Practices for use of High-Pressure Waterjetting Equipment* publication. See Table 1 An effort has been made to identify some of the technical advancements which made these pressure plateaus possible.

PSI	BAR	Color
10,000	690	Yellow
15,000	1,034	Green
20,000	1,379	Blue
40,000	2,758	Orange
55,000	3,792	Red

Table 1 - WJTA Color Code Scheme

Many of the high-pressure pumps used in waterblasting were initially driven by the needs of the oil industry. Early models were used for crude oil gathering, artificial lift, pipeline applications, drilling fluid circulation, and others. The close ties between high pressure pumps and the needs of the oil industry led to an early adoption of waterblasting as a technique used in refinery cleaning. Arguably, industrial cleaning with waterblasting started in earnest with two entrepreneurs, John Goss and Dick Paseman. These two gentlemen had enough experience to develop and copyright a pumping unit called a "Water Blaster" which was marketed by their company, American Powerstage, during the early 1960's. (Gracey,

2006) The reciprocating power pumps used for industrial cleaning during these early times were grossly unreliable with pressures above 5000psi. One weakness in maintaining the 5000psi + (plus) pressure consistently was the big block style fluid ends. As pressures increased higher loads were placed on all internal components of the pump and advancements in one area highlighted weaknesses in others.

By the 1970's, enlightened cleaning contractors realized that downtime for waterblast equipment maintenance was rather costly for asset owners. Valve components and packing elements are particularly vulnerable to wear. (Summers, 1995) As a result, Fluid Ends became the focus of many high-pressure pump developers. Initial improvements incorporated the use of cartridge style valves, use of newer alloys and stainless steels, and reducing the size of pressure chambers. Several waterblast equipment specialists started developing their own fluid end designs capable of sustaining the higher pressures, while relying on long established pump manufacturers for the powerend. John Bean, Union, and Wheatly were some of the long established powerends adaptable for Fluid End improvements by high pressure waterblasting equipment suppliers & specialists. The focus of these reciprocating power pump manufacturers was not necessarily on the needs of the waterblasting industry because their customer base was far more varied and diverse than the specific needs of a small industry.

In the late 70's and early 80's, Hydro-Services was one of the first high pressure innovators to have a resonating effect in the waterblasting industry. By adapting a Wheatly powerend to their own in-house fluid end, they were able to establish a more reliable, maintenance friendly pump for their waterblasting services. A key development in their new fluid end was their Dura-Pak, cartridge style valve. See Figure 1. (Gracey, 2006) This new valve with the corresponding fluid end enhanced the company's ability to maintain pressures of up to 10,000psi.

Figure 1 - Valve cartridge design developed by Hydro-Services

Perhaps the largest contribution to waterblasting pumps being able to reach even higher pressures of 15,000 and then 20,000psi consistently was the development of the inline valve. (Gracey, 2006) Aero-Dyne and Reliable were both leaders in developing valves that took advantage of material advancements and coaxial valve/plunger configurations. This design aligned the plungers with the valves, eliminating the need for the customary pressure cylinder. Further, the inline valve withstood higher pressures with greater consistently because the bore of the valve was no longer at a right angle to the bore of the plunger, thus eliminating the stress concentrations at the apex between the right-angle bores. This coaxial arrangement was significantly less costly to manufacture, and greatly simplified Fluid End configurations.

American Aero developed a fluid end using what was called a "racetrack" valve cartridge. The design was very similar to the in-line valve arrangement and placed the suction and discharge valve between the manifold and the stuffing box. This moved the area of fluctuating high pressure from the manifold and extended its life tremendously. The suction and discharge seats were both located on a single oval piece of metal. Though not a true self-contained cartridge, this arrangement made valve replacement easier than other designs. Another innovation was to recess the packing completely inside the gland nut instead of in the stuffing box bore. This meant that the packing came out with the gland nut instead of needing to be forced out of the stuffing box manually.

Jetstream was another company that took advantage of an existing powerend and adapted the UNx fluid end to accommodate their customer's needs. The introduction of the UNx Fluid End with the Union powerend took advantage of previously mentioned fluid end innovations. The use of a valve cartridge was used in conjunction with the inline valve arrangement but went further in incorporating both the suction valve and the discharge valve into a single cartridge, called the Univalve. See Item 57/58 in Figure 2. The design also allowed quick access to both the packing elements (Item 42) and the valve by allowing the manifold to pivot at the bottom of the rod box. Having quick access to both high maintenance items in the fluid end was instrumental in the commercial success of the Jetstream design. (USA Patent No. 4878815, 1989)

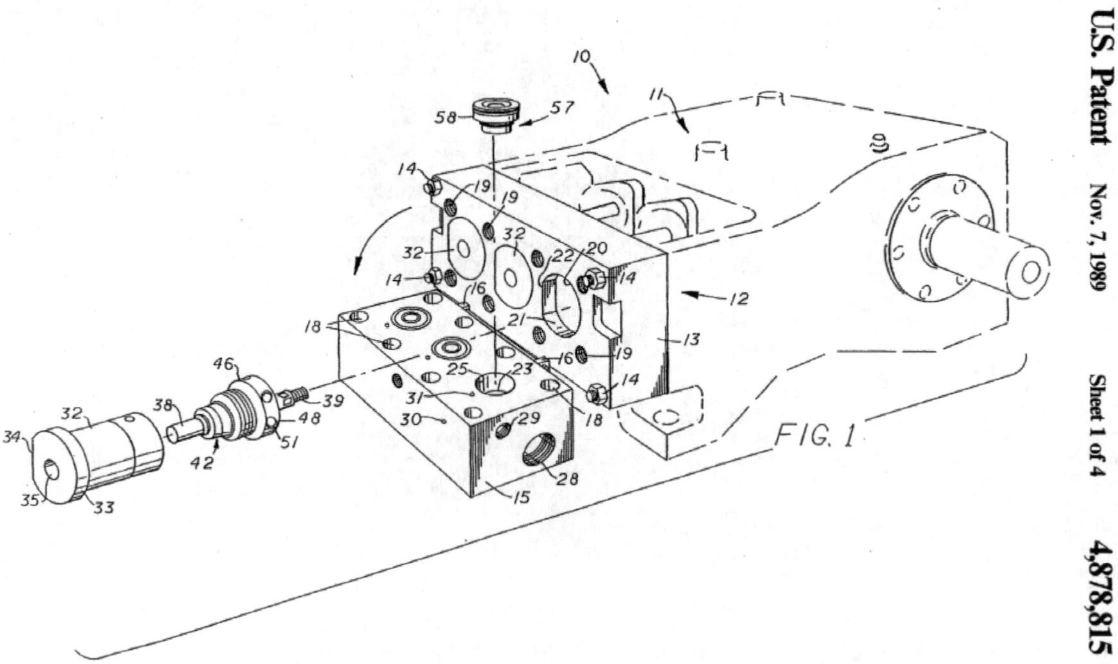

Figure 2 - UNx Fluid End

While valve improvements were continuing to advance, simultaneous advances were being made with packing seals. Of all the expendable wear items that construct a reciprocating power pump, the packing elements are the most prone to damage and replacement. Initial designs for reciprocating pumps were of the form of braided cotton rope, then interlocking square braiding with more advanced materials, such as Nylon fabric. The square cross section was formed into a ring and during operation, the sharp 90-degree corners deformed into the inner and outer diameter gaps between the plunger and the stuffing box, respectively.

Universal Packing & Gasket, now called UTEX, was instrumental in the early development of high-pressure reciprocating pump packing seals. By the mid-1960's, the benefits of stacked "V" ring seals were understood, but the use of a back-up ring made of harder material to prevent extrusion was first starting to become known. (USA Patent No. 3094337, 1963) In addition, the company developed some of the first lip seals for use with reciprocating power pumps, called the J-Design 838. (Anonymous, 2023) At the time, these seals were classified as noncrushable or nonadjustable packing. The design was unique because it was a molded ring of a Nylon impregnated with a nitrile elastomer that extended radially, both inward and outward. This configuration took advantage of the sealing characteristic known about the V-ring even further. See Figure 3(USA Patent No. 3120960, 1964) The design allowed an axial force imposed on the sealing ring to have minimal effect on the two sealing lips. The sealing ring could be clamped tightly without effecting the sealing force between inner lip and the plunger. This allowed the stuffing box gland nut to stay tight at all times without adjustment. (Henshaw, 1987) UTEX continues to develop reciprocating pump packing seals with some of the latest advancements continuing to build on the original J-Design 838.

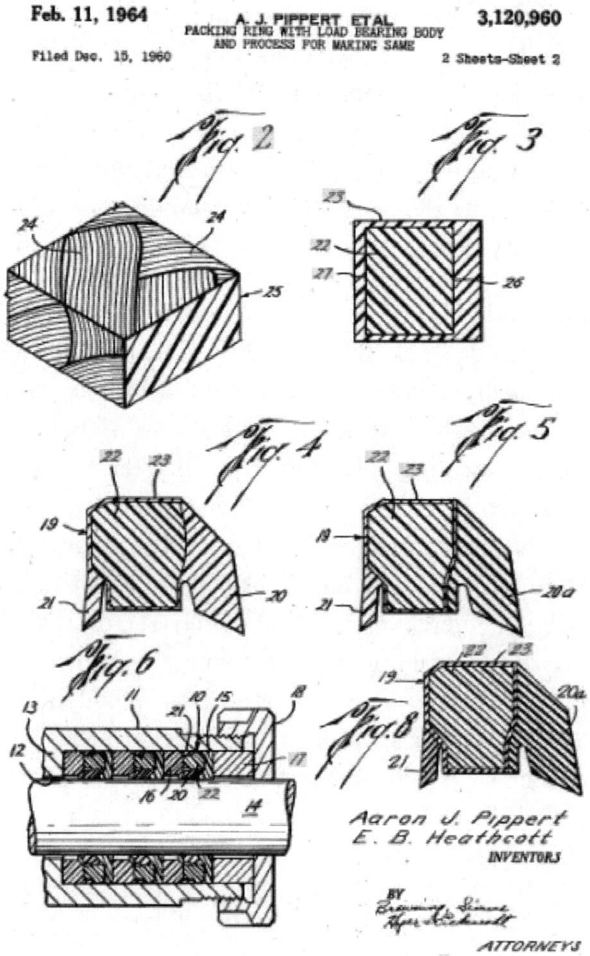

Figure 3 - UTEX Lip Seal

Plungers are yet another high wear item due to the sealing friction of the packing. Plungers can be made of ceramic, tungsten carbide, or a litany of other base materials and coatings that provide a smooth, hard surface for the packing to seal against. Ceramics have the greater performance advantage for use as plungers because they are corrosion resistant, have high hardness and can be polished very smooth. However, ceramics are brittle and costly, therefore many operators prefer tougher, lower range performance at a lower price.

Similar to plungers, many powerend advancements for reciprocating pumps were more adaptive than developmental. As the demands grew for delivering greater volume at plateau pressures, so changed the metallurgy of the rods and crankshafts to take the higher loads. Earlier metallurgical advancements, particularly in internal combustion engines, allowed pump manufacturers to adapt existing technologies such as higher strength alloy steels, forgings, and advanced heat treatments rather than having to develop the technologies themselves. However, there are certainly unique features about reciprocating pumps, such as the variable speed of the crossheads, pony rods, and plungers. Their speed at mid-stroke is greater than their speed at the ends of the stroke and this phenomenon causes pressure pulsations in the fluid called acceleration head. (Henshaw, 1987) The severity of pressure pulsations is dependent upon system variables, such as flow friction, which itself is dependent upon several factors. Perhaps this is one reason waterblasting equipment suppliers generally provide complete units to their customers.

Pumps paired with engines offer specific drive ratios, sufficient suction pressure, and take full advantage of a complete system design to optimize pump performance. Other challenges, such as lubrication constraints, pump over-speeding, and crankshaft side loading are constant reminders that pump applications in waterblasting are unique and allowed for specialization. A good example of this trend in specialization was the first 40,000psi pumping unit, offered by NLB in 1985. (NLB Corp, 2023) The launch of NLB's 40K unit set a new pressure plateau for others to follow.

New Jet Technologies shadowed NLB using a Hughes powerend. They were able to design a fluid end in the 125hp range that developed 4.2gpm at 40K in 1995-1996. (Gracey, 2006) Flow continued to increase as the company focused on increasing the speed of the pump to 750rpm. New Jet's design was unique by developing a reduction drive that allowed the capability to bolt directly on to an engine. The new drive allowed a smaller unit footprint with fixed alignment. Further, they used a pressurized and filtered lubrication system that extended the life of bearings.

In 1999, Jetstream also developed 40K fluid end adaptable to the original UNx technology. The most significant change was the reconfiguration of the valve cartridge, which put the suction valve outside the discharge valve. Similar to New Jet, Jetstream initially developed several fluid end conversions with existing brands of powerends. The success of these fluid end conversions provided a segue for Jetstream's own internally developed powerend. The first offered a three-inch stroke, the 3000, quickly followed by the 4200, with a 4.2-inch stroke. Both pumps are still offered, along with three addition pumps ranging from 75-700hp, in plateau pressures of 15,000, 20,000, and 40,000psi.

Hammelmann was the first company to offer a reciprocating plunger pump capable of sustained pressures of 3,800 bar. (55,000psi) The first Hammelmann pump to offer this pressure was well over a decade ago. Today, many of their pumps in the HAMPRO series offer operating pressures of up to 4500 bar (65,267psi). (Hammelmann, 2023)

While there has been continuous growth in waterblasting applications, many equipment suppliers do not encounter enough commercial opportunity to continue developing pumps into the 55K pressure realm. In fact, most equipment today is sold between 15K and 20K, so even the 40K pressures have limited applications. Still, higher pressures have always been at the forefront of new equipment offerings. Applications always develop after the advancements in technology that provides an ability for pumps to function at higher pressure with greater flow. Market growth for 55K plus pressure applications will continue in the future, most likely in ways unforeseen.

Works Cited

Aaron J Pippert, E. B. (1964). *USA Patent No. 3120960.*

Anonymous, A. (2023, July 09). *Wikipedia/UTEX_Industries*. Retrieved from Wikipedia: https://en.wikipedia.org/wiki/UTEX_Industries

Gracey, M. T. (2006). *High Pressure Pumps.* Houston: Oxford Press.

Hammelmann, G. (2023, May 20). Retrieved from www.hammelmann.com: https://www.hammelmann.com/en/products/pumps-and-units/overview.php

Henshaw, T. L. (1987). *Reciprocating Pumps.* New York: Van Nostrand Reinhold.

NLB Corp. (2023, October 19). Retrieved from High Pressure Water Systems & Blasting Equipment | NLB Corp: https://www.nlbcorp.com/

Pippert, A. J. (1963). *USA Patent No. 3094337.*

Stachowiak, J. E. (1989). *USA Patent No. 4878815.*

Summers, D. (1995). *Waterjetting Technolgy.* Boca Raton: Taylor & Francis.

DEVELOPMENT OF FRAMEWORK FOR THE MANAGEMENT, HANDLING SYSTEM AND MAINTENANCE IN USING HYDROSURGERY AT HOSPITALS

Mohamed Hassan Attia El-Sayed, Tasnim Firdaus Ariff and Muhammad Mukhtar Awalludin

Department of Manufacturing and Materials Engineering
Faculty of Engineering, International Islamic University Malaysia
Kuala Lumpur, Malaysia

ABSTRACT

Waterjet machining (WJM) is an advanced material removal process where the material is removed by high velocity stream of water either with or without abrasive mixture depending on its specific application. Waterjet technology has advanced into the medical and health industries and slowly becoming popular for various surgical procedures. faster healing time, less bleeding, and prevention of thermo-mechanical damage, a waterjet is always sharp and clean are seen as advantages of using this technology. Many hospitals have found that using hydro surgery is technically, practically and economically feasible. Therefore, a well-structured routine for management and handling system, and maintenance were developed through an enhanced and comprehensive framework.

Keywords: waterjet machining, framework, management and handling system, maintenance

1. INTRODUCTION

In manufacturing operations, advanced machining techniques. water jet cutting has demonstrated to be a viable technology for machining a variety of materials. It is one of the fastest-evolving machining processes, and its application in the engineering sector changes and advances every year. Additionally, water jet technology is used in the medical industry for surgical treatments, wound cleaning, and dental procedures. Over time, water jet techniques have developed into cutting-edge tools in many different types of surgery. It can be used to carefully cut skin for any kind of surgery. The pressure and cut would simply be applied by moving the tool in a straight line. The primary benefit of a waterjet incision is its accuracy; it is just as efficient as a laser cutter. However, because of the waterjet's ability to cool, the split tissue is not thermally harmed by the incision's use.

Water jet technology can be used to resect tissues, tumours, the liver, and the kidney, clean traumatic wounds, release fat cells, save nerves and blood vessels, and remove dead skin in dermatology, cholecystectomy surgery, neurosurgery, ophthalmology, oncology, and dental applications [1]. Table 1 summarises some of the medicinal applications of waterjet cutting.

Overview of using waterjet in medicine [2]
Table 1

Type of Surgery	Operation Description	Benefits
Orthopaedic	Cutting endoprosthesis and bone	Below the critical temperature by cutting
Dental	Cutting and grinding of dental materials	Reduces the risk of jagged teeth and reduces the need for anesthesia
General	Resection of soft tissues: liver, gall bladder, brain, kidney, prostate, cleaning wounds	Blood vessels and nerve fibers remain in the defined pressure maintained, minimal bleeding, intact edges and precise cuts, lack of necrotic edge, reduce the duration of myocardial ischemia
Plastic	Cleaning skin grafts, removal of tattoos, liposuction	Separation of the layers of tissue, higher accuracy of results without edema and contour changes
Dermatology	Removing dead skin	Possibility of direct dose medications in a water jet
Oncology	Rectal cancer surgery with rectal dissection	Maximum radicality and excellent autonomic nerve preservation can be achieved with acceptable postoperative morbidity and low mortality

It is important to have a systematic framework in place for the maintenance of hydro surgery equipment in hospitals in order to ensure the equipment's proper operation and longevity. This can be done by developing a framework model for a suitable management and handling system for waterjet technology in surgical purposes. A well-structured framework model for a suitable management and handling system, as well as maintenance for the use of waterjet technology in hospitals is developed to ensure that the technical, practical and economic feasibility for hydro surgery is well established.

2. LITERATURE REVIEW

Waterjet cutting also known as hydro cutting, is an industrial process by which manufacturers can cut materials. The process uses a high-pressure stream of water [3]. A high-pressure water pump pressurises the water. This water flows through high-pressure tubing into the cutting head. In the cutting head, the water flows through a nozzle, turning it into an extremely fine stream. This stream cuts whatever material is placed in front of it [4]. Manufacturing has undergone a revolution since the 1940s, allowing manufacturers to once again meet the demands imposed on them by ever-more intricate designs and durability; nevertheless, in many cases, they were composed of materials that were nearly impossible to machine. This manufacturing revolution, like past ones, is currently centered on the utilization of new tools and energy sources. This has resulted in the introduction of new manufacturing processes used for material removal known today as advanced machining processes [5].

Machining is widely utilized in health care and is crucial to the procedure's efficacy and result. Several hydo surgery applications are extensively available, particularly in orthopaedic surgery, plastic surgery, neurology, dermatology, urology, and dental surgery. Dental drilling and root canal procedures necessitate significant milling, grinding, and drilling. Various Advanced Machining processes specialize in the production of medical devices, components, medical equipment, and implants for all parts of the profession, including diagnostics, therapy, and surgery [2]. Waterjet technology has recently been developed for use in medical applications, such as soft tissue resection, bone cutting, wound debridement, and surgery [6].

3. DEVELOPMENT OF FRAMEWORK FOR IMPROVED MANAGEMENT, HANDLING SYSTEMS AND MAINTENANCE

Developing a framework model for a suitable management and handling system for waterjet technology in surgical purposes involves a systematic approach that takes into account various factors:

1. Identifying the Objectives: The management and handling system for waterjet technology in surgical applications has clearly defined objectives. This may involve increasing surgical efficiency, lowering complications, increasing safety, maximizing resource use, and developing a more appropriate and well-structured maintenance and handling system.

2. Needs Assessment: A detailed needs assessment is undertaken to understand the present issues, requirements, and gaps in the current management and handling of waterjet technology in surgical settings. Consultation with important stakeholders such as surgeons, nurses, biomedical engineers, and hospital administrators is part of this assessment.

3. Reviewing Existing Guidelines and Regulations: The existing guidelines, standards, and regulations related to waterjet technology in surgical applications are familiarized. This include guidelines from professional bodies, regulatory authorities, and industry best practices.

4. Establishing a Multidisciplinary Team: A multidisciplinary team is formed comprised of healthcare professionals, engineers, researchers, and industry specialists. This team brings a variety of viewpoints and experience to the table in order to properly construct the framework model.

5. Defining System Requirements: The specific requirements for the management and handling system are defined based on the needs assessment and regulatory norms. Safety protocols, equipment maintenance, training processes, documentation, and quality assurance techniques are all taken into account.

6. Framework Model Design: A conceptual framework model is developed that specifies the major components, activities, and interactions of the management and handling system. This model should address the identified criteria while also aligning with the objectives defined in the initial stage.

7. Incorporating Risk Management: Risk management principles are incorporated into the framework model. Determine potential dangers connected with waterjet technology in surgical applications and develop mitigation techniques. This could include putting in place safety checks, creating standard operating procedures, and conducting frequent audits.

8. Documentation and Training: Develop comprehensive documentation and training materials for the management and handling system. User manuals, standard operating procedures, training modules, and protocols should all be included. All relevant parties were assured that they would receive adequate training on the system's deployment and use.

9. Continuous Monitoring and Improvement: Mechanisms are built for continual monitoring and assessment of the management and handling system. Performance indicators are monitored, user input is gathered, and opportunities for further optimization are recognized. The framework model, which is based on fresh research, technological improvements, and changing regulatory needs, must be updated on a regular basis.

4. RESULTS

4.1 FRAMEWORK FOR MANAGEMENT AND HANDLING SYSTEMS

Developing a framework model for a suitable management and handling system for waterjet technology in surgical purposes involves a systematic approach that takes into account all necessary and relevant factors. The framework for the management and handling system is shown in Figure 1.

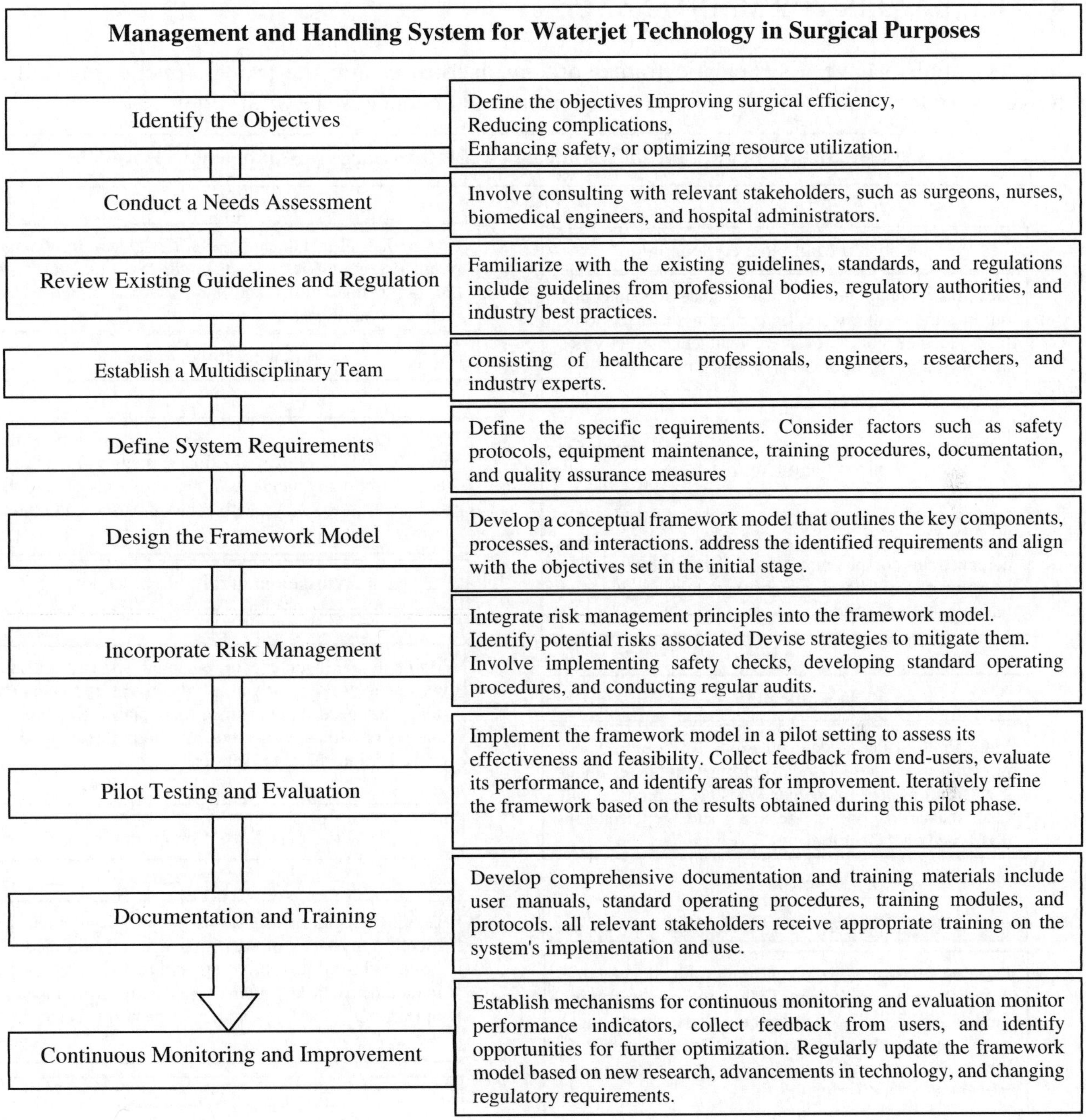

Framework for Management and Handling System for Waterjet Technology in Surgical Purposes
Figure 1

4.2 FRAMEWORK FOR MAINTENANCE

It is essential to have a systematic framework in place to ensure the proper functioning and longevity of the equipment. The framework for the maintenance is shown in Figure 2.

A systematic Framework for the Maintenance of Hydro Surgery Equipment in Hospitals

Establish a Maintenance Schedule

Develop a comprehensive maintenance schedule that outlines the regular inspection, cleaning, and servicing tasks. include both preventive maintenance activities and any required calibrations or adjustments.

Training and Competency

Ensure that the hospital staff responsible for maintaining and receives proper training and demonstrates competency in handling and servicing the equipment. cover basic operation, troubleshooting, and maintenance procedures

Documentation

Maintain detailed documentation for each piece, including manufacturer information, technical specifications, and maintenance history. a record of all maintenance activities, repairs, and replacements performed on the equipment.

Cleaning and Disinfection

Develop a standardized cleaning and disinfection protocol for the equipment. Follow the manufacturer's guidelines for cleaning agents, disinfectants, and recommended cleaning frequencies. Ensure that staff members are trained in proper cleaning procedures to prevent cross-contamination.

Calibration and Verification

Ensure that the equipment is calibrated and verified periodically to maintain accuracy and precision. Follow the manufacturer's instructions for calibration procedures or consult with qualified service technicians.

Preventive Maintenance

Conduct regular inspections of the equipment to identify any signs of wear, damage, or malfunction. Follow the manufacturer's guidelines for preventive maintenance tasks, such as replacing consumables, lubricating parts, and inspecting electrical connections.

Spare Parts Inventory

Maintain an inventory of essential spare parts and consumables. Regularly assess the inventory to identify items that need replenishment or replacement. Ensure that spare parts are sourced from authorized suppliers or the equipment manufacturer.

Risk Assessment and Incident Reporting

Perform regular risk assessments to identify potential hazards and develop mitigation strategies. Encourage staff to report any incidents or near misses related to the equipment to enable proactive measures and continuous improvement.

Compliance with Regulations

Ensure that all maintenance activities adhere to relevant regulatory guidelines and standards. Stay updated with the latest regulations and industry best practices to maintain compliance.

Framework of the maintenance of hydro surgery equipment in hospitals
Figure 2

5. CONCLUSION

The primary goal of this research was to create a thorough and systematic framework that can be used in hospitals that determine the practicality of using water jet technology into surgical procedures. By conducting a needs assessment, reviewing existing guidelines and regulations, forming a multidisciplinary team, defining system requirements, and creating the framework model, including risk management, pilot testing, evaluation, documentation, and training, as well as continuous monitoring and improvement, a well-structured framework model for an appropriate management and handling system for waterjet technology application in hospitals is developed. A well-structured framework model is created for adequate maintenance; this is required to maintain the correct operation and lifetime of the equipment by creating a maintenance schedule, as well as training and competency, documentation, preventive maintenance, calibration and verification, cleaning and disinfection, and spare parts. Inventory, vendor support, and service contracts, as well as risk assessment, incident reporting, and regulatory compliance, are all included.

6. ACKNOWLEDGEMENT

Appreciation to the Water Jet Technology Association (WJTA) for the scholarship award.

7. REFERENCES

[1] Bahls, T., Fröhlich, F., Hellings, A., Deutschmann, B., & Albu-Schaffer, A. (2017). Extending the Capability of Using a Waterjet in Surgical Interventions by the Use of Robotics. *IEEE Transactions on Biomedical Engineering*, *64*(2), 284–294. https://doi.org/10.1109/tbme.2016.2553720

[2] Hreha, P., Hloch, S., Magurová, D., Valíček, J., Kozak, D., Harničárová, M., & Rakin, M. (2010). WATER JET TECHNOLOGY USED IN MEDICINE. *Tehnicki Vjesnik-technical Gazette*, *17*(2), 237–240. https://hrcak.srce.hr/55491

[3] IQS Directory. (n.d.-b). Water Jet Cutting Manufacturers, Suppliers and Industry Information. https://www.waterjet-cutting.com/

[4] Sild, S. (2022). Waterjet Cutting – Process, Benefits and Materials Explained. *Fractory*. https://fractory.com/waterjet-cutting-explained

[5] Liang, S. Y., & Shih, A. J. (2016). Analysis of Machining and Machine Tools. In *Springer eBooks*. https://doi.org/10.1007/978-1-4899-7645-1

[6] Babaiasl, M., Yang, F., Chen, Y. S., Ding, J., & Swensen, J. P. (2019). *Predicting Depth of Cut of Water-jet in Soft Tissue Simulants based on Finite Element Analysis with the Application to Fracture-directed Water-jet Steerable Needles.* https://doi.org/10.1109/ismr.2019.8710183

2023 WJTA Conference and Expo

October 30-November 1, 2023 • New Orleans, Louisiana

Paper

REVIEW OF DEVELOPMENT OF PROFILE BY WATER JETTING FOR COATINGS REMOVAL

Lydia M. Frenzel, Ph.D.
Advisory Council
lydiafrenzel@outlook.com
Vancouver, WA USA

ABSTRACT

In the context of coatings and corrosion removal, the paint industry has held to the notion that water jetting will not change the original profile which was produced by abrasive blast cleaning. Water Jetting methods can produce a profile, generally on metals softer than steel. To make a uniform texture, the systems must keep the cleaning head with multiple nozzles moving across the surface. This paper examines the language on the current standards, the examples where coatings are removed to reveal the underlying profile, where defects in the substrate metal could occur, and the findings of a recent military paper which is limited in distribution where the researchers are concerned with the removal rate of the metal substrate when they allow the rotating deck blaster to remain stationary.

Organized and Sponsored by WJTA

1. INTRODUCTION

The author started this paper thinking that the Naval Surface Warfare Center Technical Report (NSWCCD-TR) with a mountain of data on the removal rate of aluminum and steel with a High Pressure Waterjetting (hpwj) crawler could be compared with prior technical data reports. See Reference 1.

This concept proved almost insurmountable because the NSWCCD TR and the prior reports came to profile formation from two different points of view- that of the practitioner with practical experience and that of the authority with overall comprehension. As an example, the Practitioner can run the gasoline engine all day and repair it while the Authority invented the gasoline engine, but not necessarily know the ins and outs of everyday usage.

The NSWCCD focused on misuse of the equipment; the prior reports focused on the positive aspects of UHPWJ enhancing adhesion, formation of a micro-profile over a larger abrasive blast profile and removing embedded abrasive. The prior reports noted the same misuse but didn't emphasize it. The profile produced by the UHP WJ was to be painted, not reconfigured by additional abrasive blasting as mandated by the Navy. However, flight deck maintenance is a safety issue and is an area of expertise in itself. Flight decks are discussed at every meeting.

Starting around the mid 1970's, the author started working with the U.S. Navy through the industrial-Navy program of the National Shipbuilding Research Program (NSRP) to get acceptance of waterjetting as an alternative surface preparation method to conventional abrasive blast cleaning. Dr. David Summers, retired University of Missouri-Rolla, and Roland Lever, Cavitech, were working with other divisions of the Navy and Department of Defense. Europe was seeking to minimize waste streams. Abrasive blasting to remove coatings, or to create the original profile, was the largest waste stream in shipyards. (See Reference 2) The author recognizes Dr. David Summers, Roland Lever, Larry Fulmer, Dr. Mohan Vijay, and the StoneAge group as people who are both Practitioners and Authorities.

Rolland Lever taught the Navy the practical aspects of WJ stripping of coatings; his company photographs and descriptions were adopted by the Navy before industry standards. See Reference 3

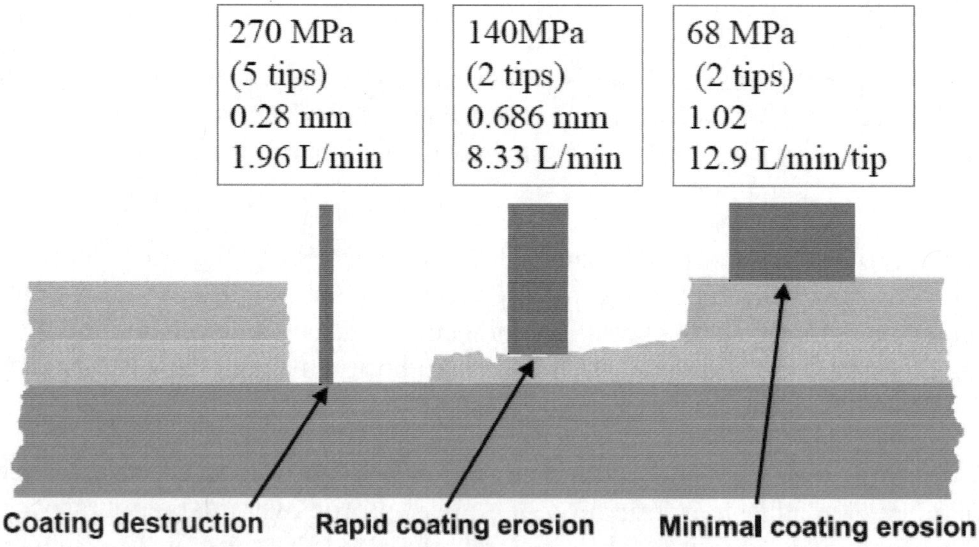

Cross Sec mm^2	0.0613	0.369	0.811
Jet Velocity m/sec	738	522	363
Impact force/tip kg	2.4	7.3	8.1
Jet Energy kJ	89	189	142
Energy Intensity kJ/mm^2	1,450	512	175

Figure 1 Summary Graph of Pressure, flow, and Removal Reference 3

Lever clearly depicts that as the pressure increases, the tip diameter decreases and the energy intensity (impact) per area increases dramatically. Lever also shows that WJ can produce profile in intermediate coatings layers.

Working with a customer can be tricky. Contractors and coatings manufacturers are reluctant to tell any customer that their processes might not be working or that a change could give better results. Criticism leads to loss of contracts and revenues. As a result, misinformation can reign supreme because practitioners see something that they don't understand from their experience, and they resist continuing education.

To this day, at least for the Navy, so far as the author can tell as an outside observer, there is a handful of Naval civilian technical researchers who have worked with water jetting via Dr. Summers, myself, and a few major contractors. However, the author has observed an absence of people who have been to a WJTA conference and understand the fundamentals. Their knowledge comes from the deck plate jetters.

In 1994, the U.S. Navy invented Removal of Coatings by Water Jetting in Bremerton Marine Coatings Conference where they demonstrated the Pratt and Whitney Ship ARMS system. Batelle was doing similar Water jetting coatings removal for the Air Force. Critical to the Batelle work was the Effect of the Water Jet stream on the aluminum substrates. Reference 2, 5

When water jetting was introduced to SSPC, NACE, and the NPCA (National Paint and Coatings Association) around 1985, the immediate response from the industry was to resist anything that was not conventional abrasive blasting, or hand-power tools. As we worked on making standard language, the task groups settled on the template language:

> "Waterjet cleaning to achieve the "Very Thorough Cleaning (WJ-2)" degree of surface cleanliness is used when the objective is to remove almost all rust and other corrosion products, coating, and mill scale, but when the extra effort required to remove all of these materials is determined to be unwarranted. Discoloration of the surface may be present.
>
> ***Waterjet cleaning does not provide the primary anchor pattern on the metallic substrate known as "surface profile."***

See Reference 6

2. WATERJET CLEANING

Over the last 40 years, the author has taken many calls for advice or education concerning water jetting, visual appearance, and the measurement of conductive ions. Now the author is getting calls "We have a proposal for the profile to be made by water jetting. Is this possible?" The answer is: "Yes, it is possible, certainly on pieces which are handled by automated systems. On a large surface or structure, it might or might not be economically viable. Contact the manufacturers of the WJ pumps as the pressure in the systems have moved from 10,000 psi (68 MPa) up to 50,000 psi (340 MPa)."

What is the "standard" language and how can it be mis-interpreted?

> ***Waterjet cleaning does not provide the primary anchor pattern on the metallic substrate known as "surface profile."***

The author was careful to let the paint industry know that WJ removed material and could produce patterns in metals, but not in an economical time frame.

However, the coatings industry adopted a black and white picture. Practitioners and the abrasive suppliers did not want the hint that WJ could supply a profile. The Practitioners placed their conservative interpretation into the NACE CIP course.

2.1 Background of NSWCCD-61-TR-2020/15

"According to the NACE® Coatings Inspector Program (CIP) Level 1 [4] training course section 10.17.4 states, "UHPWJ will not produce a surface profile but **can restore** any previously existing surface profile if the equipment is designed to clean the surface to a high standard." This study will help determine if this statement holds true for UHPWJ operations to remove nonskid onboard naval flight decks due to the frequency at which flight decks are replaced over the service life of a ship." See Reference 2, p.3.

The above statement can lead practitioner to an erroneous position. It is true that you can remove coatings without substantially changing the underlying substrate. It is also true that you can change the substrate energy and profile, typically to form a micro-profile rather than the macro profile.

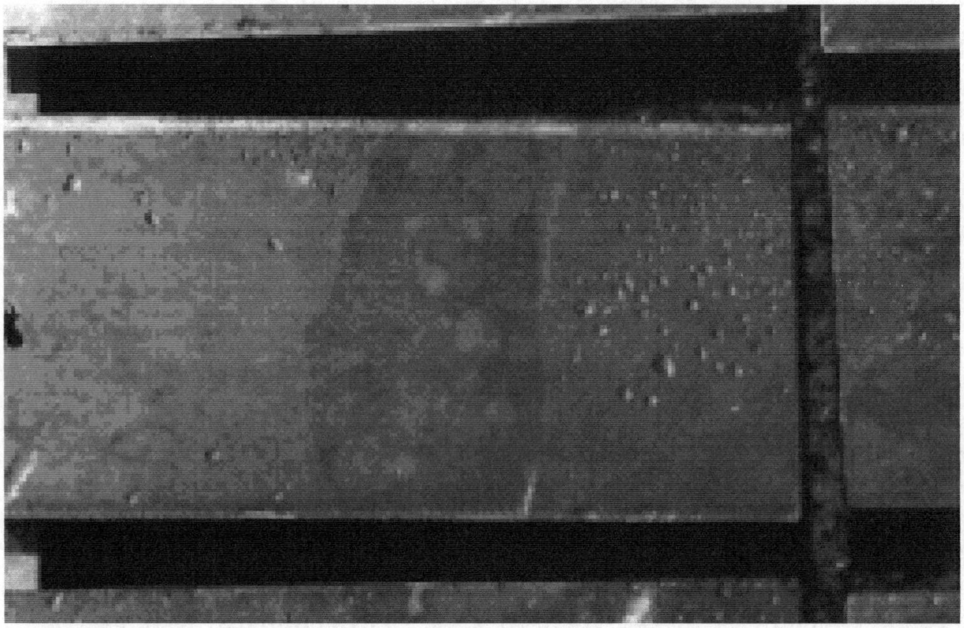

Figure 2 Example of water wetting the surface better after water jetting. The center was water jetted at 40,000 psi with 6 gpm. Left and right of center were only degreased. Reference 7

In 2001, the Navy was investigating the formation of surface profile at the Naval Sea Systems Command. NSTM Chapter 634, Deck Covering requires a measurable anchor tooth profile of 3 to 4.5 mils. Their conclusion. Water Jetting at pressures up to 40,000 psi alone cannot impart this profile [on cold rolled steel]. It can only reveal an existing profile. During surface preparation, the wetting characteristics of the surface changed significantly. Prior to waterjetting, the water was observed to bead on the surface meaning incomplete wetting was achieved and the substrate had low surface energy. After the surface was waterjetted, however, water was observed to spread quickly over the surface, meaning more complete wetting was achieved, thus the surface energy of the substrate was increased.

- The profile data on the solvent cleaned non-blasted panel and the waterjet cleaned panel were similar. The effects of water jetting on surface profile were undetectable using either Testex tape or profilometer.
- The data show no significant difference between the adhesion of coatings to the various surface preparations used.
- Water jetting may affect substrate surface energy.
- The results reported in this paper are based on the specific parameters of the water jetting equipment used, substrate and coating materials selected, and the tests performed.

McGaulley is requesting additional panel testing and determination on other factors that affect adhesion. From this project, the Navy could conclude that WJ alone would not create a profile. See Reference 7

3. CONVENTIONAL PHOTOS

The following photos illustrate what a practitioner would encounter and why the revealed substrate looks like the original profile under the paint or rust.

Figure 3 Waterjet cleaning to WJ-1 revealing the abrasive blasted substrate. No measurable change in profile. The two layer coating system in white topcoat, red primer.

Figure 4 Waterjet cleaning to WJ-1 of Grade D steel. The rotating head standoff was moved further in. Showing the linear groove of the nozzle.

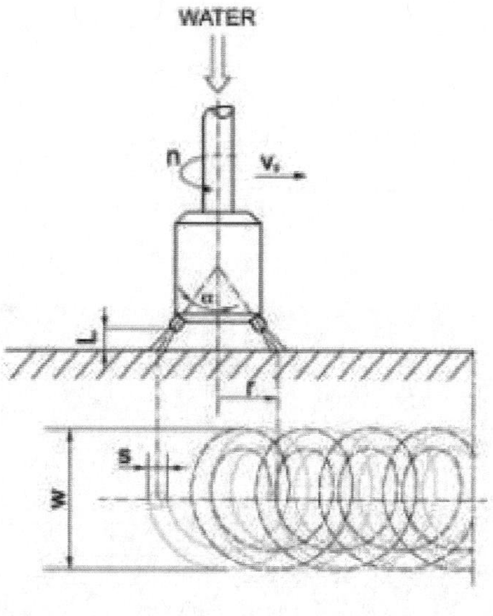

Figure 5 Flat Surface Treatment Using Rotary Head Equipped with two water nozzles Reference 8

Kunaporn and Ramulu published several papers on the impact pressure and peening and changing the surface energy of various substrates. Water Jet cleaning can influence the energy of the substrate surface and enhance fatigue strength. See Reference 9

The Navy wasn't alone in searching for the effect of water jetting on the substrate profile. Akzo Nobel had just issued their Hydroblasting photos in 1994. Europe was adopting this technique. Individual coatings manufacturers were doing their own proprietary research on specific coatings systems, primarily for marine exposure.

In 1995 as WJ for coating removal was being introduced to SSPC, Stein Bjoerndal, Ameron International- Norway, said that Ameron Norway went over steel with UHPWJ with 10 passes; they found a 10% decrease after 10 passes. This project was reported in a Norwegian Marine Tanker Coatings magazine, but translation from Norwegian to English was not available. The report has been mislaid during the interim years. Ameron wanted to verify, in their in-house R & D, the influence of Water Jetting on the profile (peak to valley). Reference 10

The author wrote about revealing the substrate in 1983 to 1985 using 140 MPa (20,000 psi) waterjetting. The author's experience at 140 MPa (20,000 psi) was that corrosion was removed and the profile remained at originally established. (Reference 11) Dupuy and Howlett published a paper for 35,000 psi (241 MPa) water blasting on surfaces which were original blasted with abrasive. (Reference 12)

It is noted that we were not looking for a change in surface profile, nor trying for a deliberate mis-use of the equipment. We were seeking the removal of rust, prior coatings, and invisible contaminants.

4. DEFECTS DESCRIBED IN OTHER PAPERS

The conclusion from all the paper- keep the head moving to avoid localized deterioration of gouges or grooves. Use a uniform motion and standoff to produce a uniform textured substrate.

4.1 Miller and Swenson

In 1999, Miller and Swenson, Thiokol Propulsion Division, was investigating erosion rates between the 10,000 to 15,000 psi (68 MPa to 100 MPa) fixed nozzle systems to 36,000 to 40,000 psi (250 to 272 MPa) multi-orifice, rotary nozzle, waterjet system. The project was to determine a method to refurbish space shuttle equipment with the least amount of erosion. This project is close to the requirement of the NSWCCD-TR. Reference 13

> Maximum erosion at any one location is dependent on the nozzle overlap pattern. Calculating erosion based on weight loss proved to be the only method sensitive enough to detect the slight erosion of D6AC steel by the established waterjet cleaning process. All erosion testing was conducted on bare metal coupons to simulate a 'worst case condition'.
>
> The project was completed at 2 different sites with slightly different paraments: Their parameters were:
> 1992
> - Water Pressure: 36,000 psi (at the pump)
> - Nozzle Speed: 1,000 rpm
> - Nozzle Standoff: 1.0"
> - Nozzle Angle: Normal to the surface being cleaned
> - Sweep Rate: 70 inches per minute
>
> 1995 and 1998
> - Water Pressure: 40,000 psi (at the pump)
> - Nozzle Speed: 500 to 1,500 rpm (1,300 rpm nominal)
> - Nozzle Standoff: 1.0" to 2.5" (2.5" nominal)
> - Nozzle Angle: Normal to the surface being cleaned
> - Sweep Rate: 30 to 60 inches per minute (60 ipm nominal)
>
> **1995 Testing**
> The 1995 testing was conducted to qualify a new automated ultra-pressure, rotary nozzle, waterjet system for use on space shuttle flight hardware. Data from the testing verified that the new system would safely and efficiently remove contaminants without damaging the hardware. This hardware was designed to be reused up to 20 times. Component wall thickness was designed with the assumption that each refurbishment would remove up to 0.001 in. of material from the steel substrate wall thickness.
>
> **1998 Testing**
> The 1998 testing qualified the same automated system for use on Minuteman III Stage 1 hardware. The task of the Minuteman Propulsion Replacement Program (PRP) is to remove old propellant, insulation, and adhesives from the steel hardware and then to reline and reload

the hardware. The Minuteman hardware was not designed for reuse and therefore has minimal allowances for erosion of the steel substrate. This testing verified that the minimum allowable erosion of 0.0001 in. would not be exceeded during the cleaning process.

Erosion testing was determined by calculating the weight difference of each test sample (before and after waterjet exposure) and then dividing by the material (D6AC steel) specific weight and coupon exposed surface area to determine the erosion (mils). In comparing the single pass waterjet erosion of 0.009 mils (0.23 μm) to that of the zirconium silicate paint removal of 0.7 mils (18 μm)., the level of material erosion is decreased by approximately 98%. See Table 1

The rate of D6AC steel erosion (inches/pass) when exposed to the ultra-pressure waterjet cleaning process is not linear. The erosion rate is only slightly different between the first and second exposures (~ 11%) while the third and subsequent exposures display a much larger variation in the erosion rate. In fact, the data show that the initial exposure removed up to 88% more material than the subsequent exposures.

Failure simulation testing shows that any prolonged exposure, at zero nozzle rpm and / or zero sweep rate, will cause significant material removal, (0.0017 in./sec.).

Miller and Swenson also illustrated the hole drilling when the head was not rotated and crop circles when the head was rotating without a transverse movement.

To prevent unacceptable erosion of steel substrates it is essential that automated systems be designed with devices to control and monitor the critical processing parameters. These systems must also be equipped with the means to automatically and immediately shut off the flow of high-pressure water when established critical parameter limits are violated.

Table 1. Normal Operating Parameters Erosion

D6AC Steel Erosion Data					
Pressure (psi) 10,000 psi = 68 MPa	Revolution/min (rpm)	Stand-off (Inch) 1" = 25 μm	Volume Liter/min (lpm)	# of passes	Average Erosion mil-inch
36,000	1,000	1 "	70	1	0.009
40,000	1,300	2.5"	60	2	0.017
40,000	1,300	2.5	60	3	0.018
40,000	500-1,500	1.0 – 2.5	30-60	6	0.018
Zirconium silicate grit				1 refurb	0.700

Note: Figure 6 in Reference 13

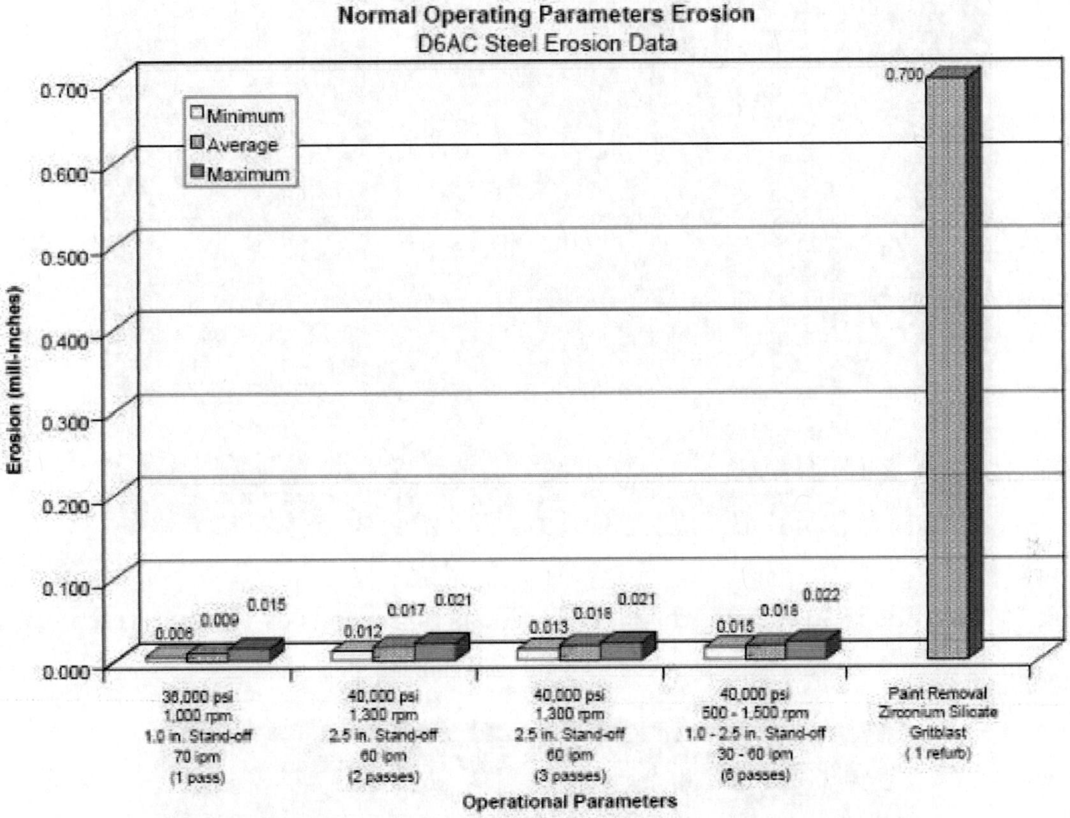

Figure 1. **Waterjet Erosion versus Gritblast Erosion**

Figure 6 Normal Operating Parameters Erosion

Note: Figure 1 in Reference 13

4.2 Wright, Wolgamott, Zink (StoneAge group)

Wright, Wolgamott, and Zink published a series showing localized distortions and what happened if the head stopped rotating or was not moved along the substrate. They used a rotating jet, pipe which is found in sewer lines, 20,000 psi (140 MPa) both good and bad nozzles, with varying dwell times. Steel process lines and tubes are commonly cleaned using waterjet systems with pressures up to 40,000 psi. There is a risk of damaging these lines depending on operating parameters such as jet pressure, angle, rotation, and rate of traverse. See Reference 14

Waterjet surface preparation is typically performed using pressures from 20,000 to 40,000 psi, with rotating nozzle heads varying in diameter from 2 inches to 16 inches. Materials being removed include coatings, oxidation, or scales. The purpose of this research was to determine the effects of variables such as standoff distance, traverse speed, surface speed, rotation speed, and the head design.

Figure 7 Left side is 30 seconds dwell, right side is 60 seconds dwell time Reference 14

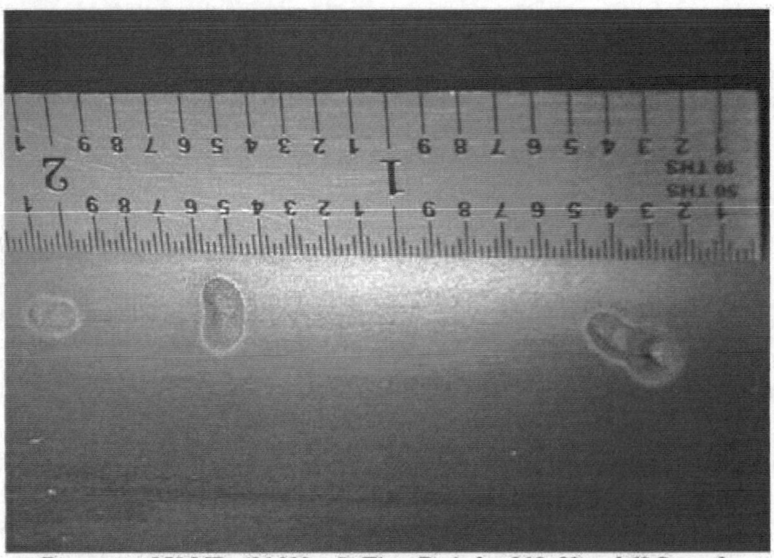

Damage at 250 MPa (36,000 psi), Time Periods of 10, 30 and 60 Seconds
Figure 9.

Figure 8 Damage at 250 MPa, time periods of 10, 30, and 60 seconds Reference 14

5. CRAWLER UNIT NSWCCD-1-TR-2020/15

In Figure 1 Lever clearly depicts that as the pressure increases, the tip diameter decreases and the energy intensity (impact) per area increases dramatically. With a robotic crawler, the operator is placing 5 to 8 tips which are about 0.025 cm diameter (0.01 inch), rotating the nozzle bar, and trying to clean a 30-60 cm wide (12 to 24 inch) path. The transverse speed must be constant, or you will make overlapping spirals. See Figure 5

The NSWCCD-TR is more comprehensive than any individual WJTA papers. Reference 1. It is unclassified, but with distribution to persons with contracts with the US Navy. Reference 13 is closest to the NSWCCD-TR.

Reference 1 is 109 pages long. Selected portions and photos are given in this paper. In this section, the author has kept the original numbering of the Figures; thus, the TR figures are out of sequence with the rest of this paper. The author has changed the original Table numbers to conform with the Table numbering in this paper. Reference 1 uses English notation.

5.1 Operating Parameters

The conditions of the US Navy system (Reference 1, p. 5) are:

- 272 MPa (40,000 psi),
- 2 passes per cycle,
- 1 ft^2/min,

The objective of this project was to determine **if there was degradation** to the abrasive blasted surface profile of steel and aluminum (Al) plates after seven repetitive cycles of nonskid coating application and removal (replacement) by ultra-high pressure waterjet (UHPWJ) using a commercial robotic crawler unit capable of pressures at or above 40 ksi.[sic 40,000 psi] Results from this study will help determine if the minimum requirement of the 20 percent abrasive blast mandate in the current FY-20 CH- 1, NAVSEA Standard Item (NSI) 009-32 is sufficient during nonskid system replacement on flight decks.

5.2 Executive Summary

The *objective of this project was to determine if there was degradation to the established surface profile of steel and aluminum (Al) plates after seven repetitive cycles of nonskid coating application and removal (replacement) by ultra-high pressure waterjet (UHPWJ) using a commercial robotic crawler unit capable of pressures at or above 40 ksi. Results from this trial study will help determine if the minimum 20 percent abrasive blast mandate in NAVSEA Standard Item (NSI) 009-32, FY 21, CH-2 is sufficient during nonskid system replacement on flight decks. Four plates were selected for this project: HY-80 (two plates), Al 5456-H116, and Al 6061-T6. An Al 6082-T5/T6 extruded section was added during testing via sponsor request. Five test methods were selected to evaluate changes to the plates' surfaces: profile measurements using a needle depth gauge, ultrasonic testing (UT), metallography, adhesion testing, and mass loss testing on Al 6082.*

Surface profile measurements using a needle depth gauge revealed that the HY80- 1 and HY80-2 steel plates resulted in an average surface profile rate of reduction of 0.21 mils/cycle and 0.29 mils/cycle, respectively. The Al 5456 and Al 6061 plates revealed an average surface profile reduction rate of 0.27 mils/cycle and 0.36 mils/ cycle, respectively. The Al 6082 extruded section revealed an average profile rate of reduction of 0.3 mils/cycle, respectively. The HY80-2 and Al 6061 plates that received 14 passes of UHPWJ after seven cycles of nonskid replacement revealed a greater rate of reduction than the HY80-1 and Al 5456 plates that received seven passes of UHPWJ by approximately 0.8 mil/cycle and 0.9 mil/cycle, respectively.

UT longitudinal wave technique revealed that the HY80 steel plates resulted in a material thickness rate of reduction after seven cycles of nonskid replacement of 0.21 mils/cycle and 0.29 mils/cycle. The Al 5456 and Al 6061 plates resulted in a material thickness rate of reduction after seven cycles of nonskid replacement of 0.36 mils/cycles and 0.43 mils/cycles. The HY80-2 plate, and Al 6061 plate that received 14 passes of UHPWJ both revealed a greater rate of reduction than the HY80-1 and Al 5456 plates in order of magnitude of 0.8 mils/cycle and 0.7 mils/cycle

Metallography images of the HY80-1 steel plate samples revealed that after six cycles of nonskid replacement the peaks became smoother. Metallography images of the HY80-2 steel plate samples that received 12 passes of UHPWJ after six cycles of nonskid replacement revealed more of a rounded surface profile with areas of little profile.

Metallography images of the Al 5456 plate samples revealed less frequent deep valleys and high peaks after six cycles of nonskid replacement, while the Al 6061 plate samples after 12 passes of UHPWJ revealed that periodic high peaks seen on the baseline profile images were generally not present.

Mass loss testing of Al 6082 resulted in a linear net increase in mass loss and material thickness loss as the dwell time of the crawler unit waterjetting at 40 ksi in a stationary position increased. While the original surface profiles for all plates decreased with increasing cycles of UHPWJ, the adhesion strength of the primer coating remained good with average pressures far above minimum requirement after six cycles of nonskid replacement.

5.3 Types of Material Used

The following list is comprised of the materials selected for this study:

- Aluminum 5456-H116 (4' x 8' x 0.25")
- Aluminum 6061-T6 (4' x 8' x 0.25")
- HY-80, (two plates total) (4' x 8' x 0.25")
- Aluminum 6082 (4' x 4' x 0.125") (Note: This was not a plate. This section was an extrusion with corrugated cross-section. The thinnest part of extrusion was 0.125" thick. The total height of extrusion was 3.5".

5.4 Surface Profile Measurement Preparation

For this study, the HY80 steel plates, Al 5456 and Al 6061 plates, and the extruded Al 6082 material were abrasive blasted once in accordance with NSI 009-32. Profile measurements were taken in 12 different spot locations in accordance with ASTM D4417 Method B [6] (DeFelsko Positector 6000 series surface profile needle depth gauge).

Note by Author. The surface profile needle depth gauge will not be sufficient to measure micro profile. It will measure maximum profile, which is appropriate for this study. NSI 009-31 is Navy Standard Item 009-32 Cleaning and Painting Requirements. It is revised periodically; Dated versions are located on the internet.

Paragraph 3.11.9.3 in the NSI 009-32 states, "If the primer coat is not overcoated with nonskid within 7 days of final full primer coat application, the primer must be removed, and the surface preparation repeated." The HY80-2 and Al 6061 plates were coated with an epoxy "hold" primer on one-half while the other half of the plate was left uncoated. Afterwards, the plates were placed outside overnight for a period of 24 hours or until flash rust and oxidation appeared on the bare section, then they were waterjetted again to remove the holding primer and oxidation prior to applying the nonskid system. This method was selected to reflect two common field practices of nonskid replacement on naval ships' flight decks where both require an additional pass of UHPWJ to reestablish the required surface cleanliness prior to nonskid system application. The HY80-2 and Al 6061 plates received 14 passes of UHPWJ in total for the seven cycles of nonskid replacement. For this study, the application and removal of the nonskid system (primer and topcoat), and removal of holding primer/oxidation via UHPWJ is considered one complete cycle. For the HY80-2 and Al 6061 plates, one cycle equaled two passes of UHPWJ. Contractors during nonskid replacement will either apply a holding primer after the old nonskid system was removed or leave the deck's substrate exposed for multiple days before coming back and waterjetting again to remove any holding primer, flash rust (steel decks), or oxidation (aluminum decks).

5.5 Observed Defects Due to UHPWJ (page 16)

The UHPWJ unit used was non-hydraulic and required one man to maneuver it back and forth. The transition rate of the unit used on the plates was recorded at 1 ft2/min. Deep gouges on the Al 5456 plate surface were observed after the UHPWJ passes; these consisted of either a linear row of eight deep gouges displayed in Figure 17 or circular grooves displayed in Figure 18. The robotic crawler unit has eight jewel-nozzles lined up on an eight-inch straight bar which rotates. The bar is mounted to the crawler unit and moves over the plate surface directing the water at ultra-high pressure. The linear row of deep gouges can occur when the ultra-high pressure water is exiting the nozzles while the bar is not rotating and the crawler unit is stationary (see Figure 17). Upon discovery of the linear defects, the operator of the robotic unit was questioned, and he confirmed the defects were created when the bar was not rotating. One set of the gouges displayed in Figure 17 were cutout from the Al 5456 plate. Cross-sections of the gouges were examined using metallography to reveal the depth of penetration made into the plate surface along with 3D heat imaging illustrated in Figures 19 and 20. In Figure 19, the depth of the gouges measured at 39.4 mils (1002 μm) and 36.6 mils (929.73 μm). In Figure 20, a 3D heat map confirmed the measurements of the metallographic cross-section images of the deep gouges to be between 37.4 – 39.4 mils (950-1000 μm).

The second type of defects as seen in Figure 18 are circular grooves (also called crop circles) that can occur during the UHPWJ process when the robotic crawler unit is left stationary but the ultra-high pressure water is running and the bar is rotating. They can also occur when the operator is maneuvering the unit back and forth in the same general area. Circular grooves visually revealed into the Al 6061 plate during the cycles of UHPWJ as illustrated in Figures 21 and 22. The circular grooves were examined in more detail during the mass loss trials of Al 6082.

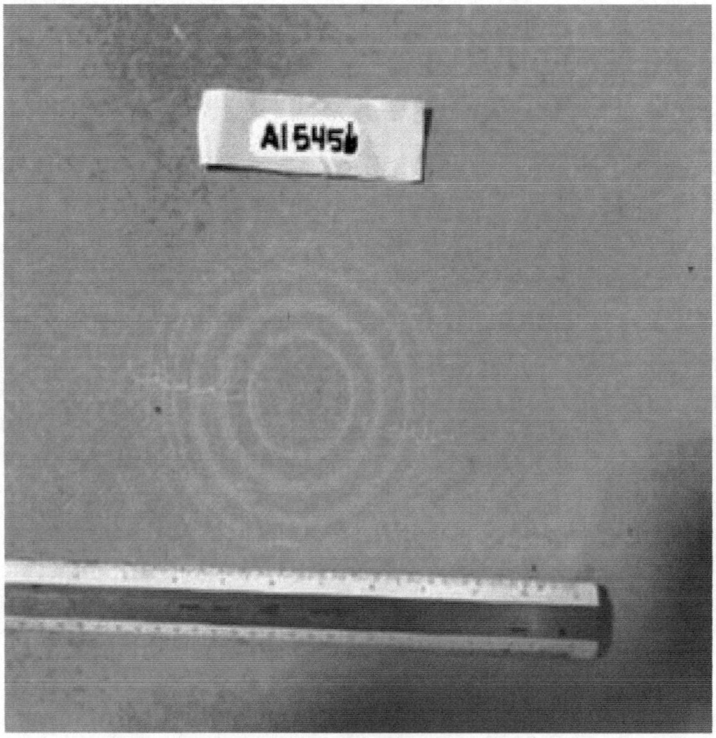

Figure 18 Combination of deep gouges in a linear direction and circular grooves seen into the Al 5456 plate. NSWCCD Figure

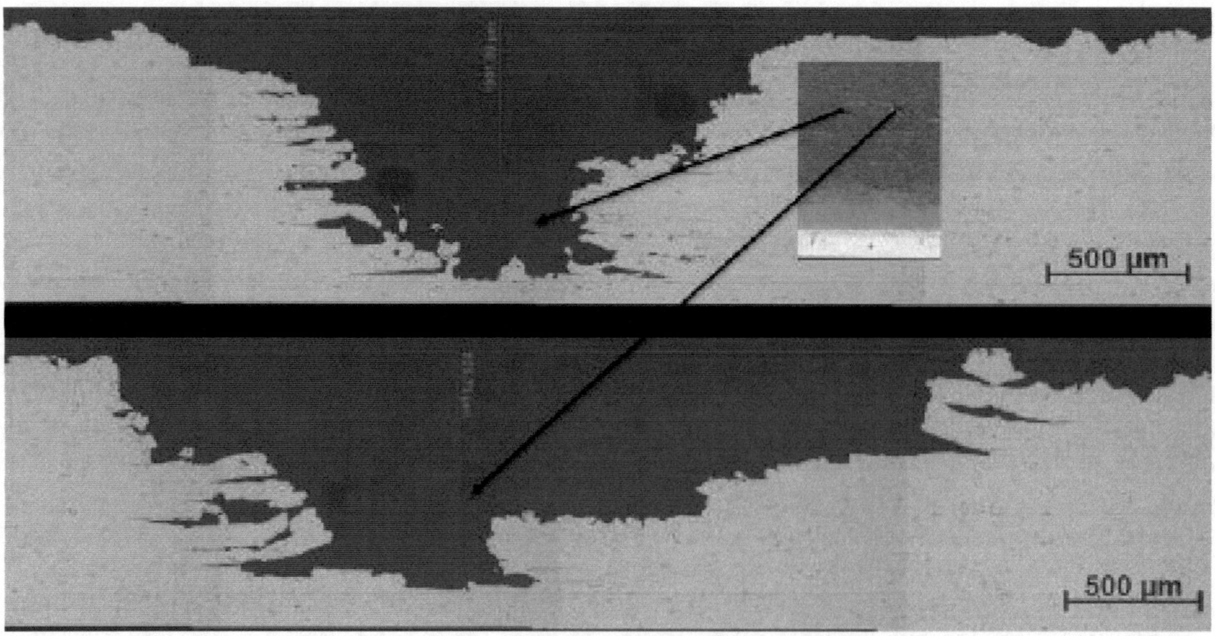

Figure 19 Metallographic cross-section images of two of the narrow gouges in the Al 5456 by a robotic crawler unit during the process of UHPWJ.

5.6 Ultrasonic Testing Longitudinal Wave Technique

This technique established the material thickness.

The HY80-1 results displayed a material thickness loss ranging from 0.001-0.0015 in. (1.0-1.5 mils) after seven cycles of nonskid replacement.

The HY80- 2 plate displayed a material thickness loss ranging from 0.0015-0.002 in. (1.5-2.0 mils) after seven cycles of nonskid replacement (14 passes of UHPWJ).

The Al 5456 plate resulted in a material thickness loss ranging from 0.001-0.0025 in. (1.0-2.5 mils) through seven cycles of nonskid replacement.

The Al 6061 plate resulted in a greater material thickness loss than the Al 5456 plate ranging from 0.0015-0.003 in. (1.5-3 mils) after seven cycles of nonskid replacement (14 passes of UHPWJ).

5.7 Surface Profile Evaluation by Metallography

Note: The text is from NSWCCD-61-TR-2020/15 and refers to the Figure Numbers as in the Technical Report. The **tables** have been renumbered within the text to reflect consecutive tables in **THIS** paper.

Representative metallography images of cross-sections from the HY80 steel panels, Al 6061 panels, and Al 5456 panels of the baseline, after the first cycle, and after the sixth cycle of nonskid replacement are shown in Figures 43 through 55. The profile images of the HY80-1 samples observed by metallographic examination are shown in Figures 43 through 45. The metallographic images of the HY80-1 samples revealed a varied range of profile where in some areas examined showed little profile such as those shown in Figures 43a, 43b, 44a, and 44b, while other areas showed large peaks such as those shown in Figures 43c, 44c, and 44d. Overall, while the large peak ranges did not change significantly after six cycles of nonskid replacement, the peaks became smoother. Metallography was not performed after the seventh cycle because the plates' surface were significantly gouged during shipment.

Figure 58, the depth of penetration of the mechanical gouge made into the coupon after UHPWJ for 10 seconds was measured at 1.32 mils (33.57 μm).

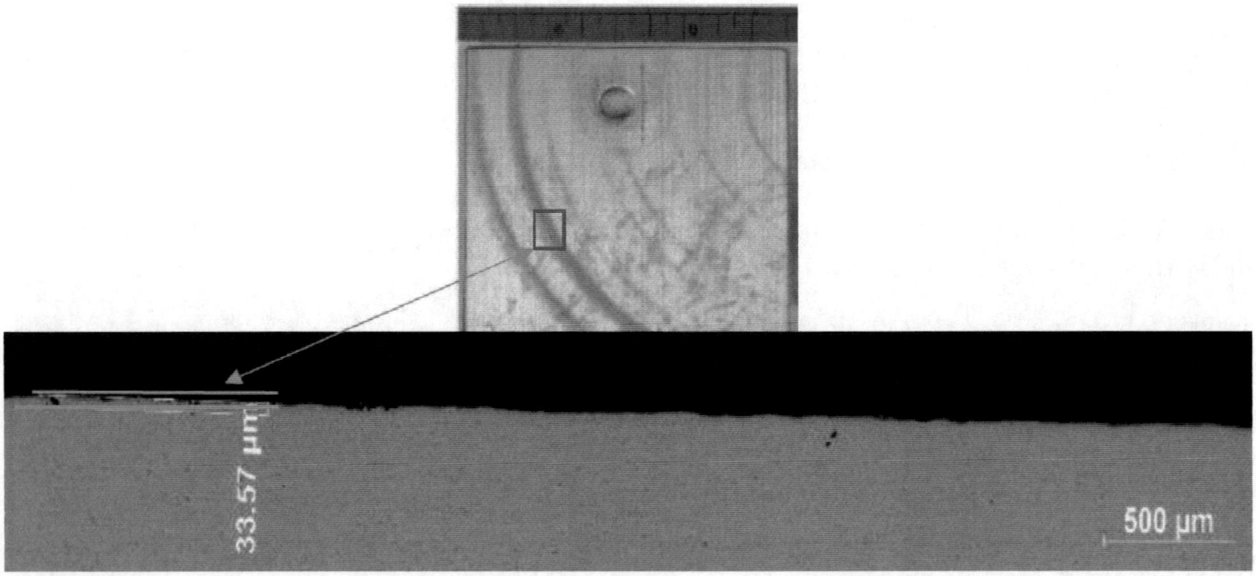

Figure 58. Metallographic image of a non-abrasive blasted Al 6082 coupon after UHPWJ for 10 seconds

Representative metallographic images of non-abrasive and abrasive blasted Al 6082 coupons after UHPWJ for 20 seconds are shown in Figures 60 and 61. In Figure 60, the mechanical gouge made into the non-abrasive blasted coupon was measured at 8.6 mils (218.03 µm). The measured depth of penetration of 8.6 mils equals 3.6% material thickness loss of a 0.236-inch thick plate. In Figure 61, the depth of penetration of the mechanical gouge made into the coupon was measured at 4.7 mils (120.67 µm). The original surface profile was 3.07 mils (77.97 µm).

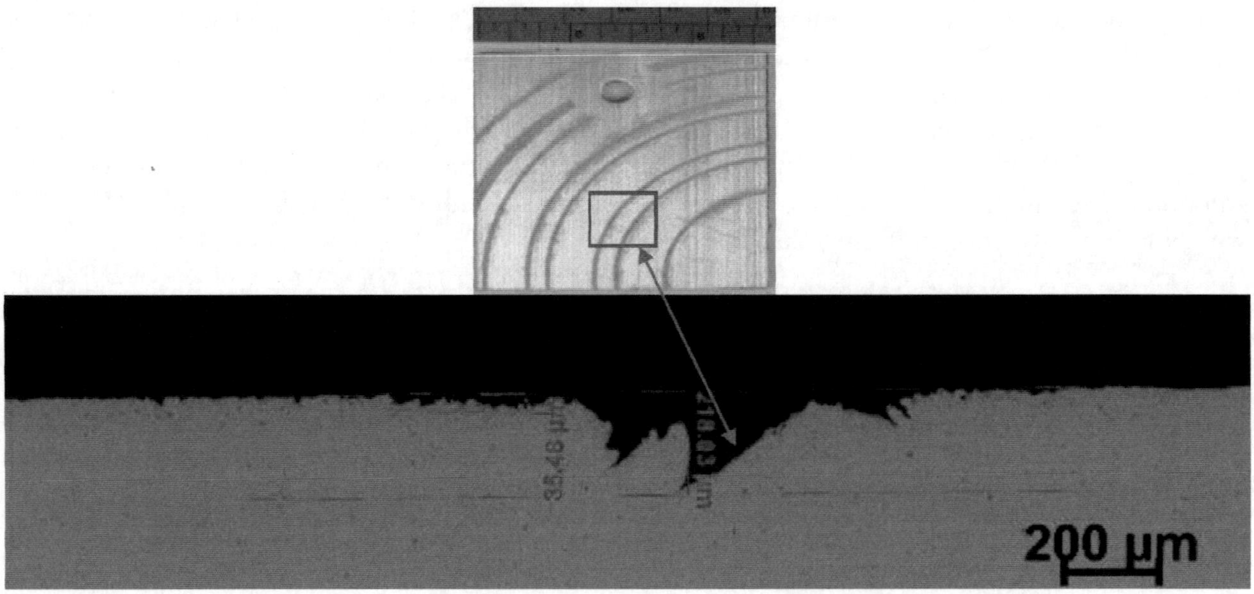

Figure 60. Metallographic image of a non-abrasive blasted Al 6082 coupon after UHPWJ for 20 seconds

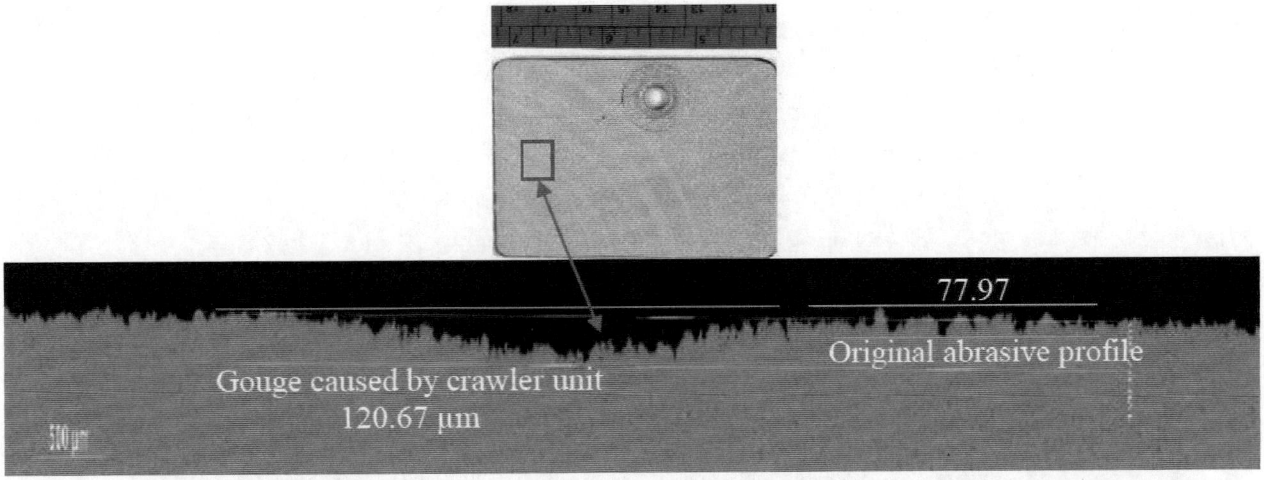

Figure 61. Metallographic images of an abrasive blasted Al 6082 coupon after UHPWJ for 20 seconds

Representative metallographic images of a non-abrasive and an abrasive blasted Al 6082 coupon after UHPWJ for 100 seconds are shown in Figures 68 and 69. In Figure 68, the mechanical gouges made after UHPWJ into the non-abrasive blasted coupon were measured at 12.4 mils (314.27 μm) and at 9.4 mils (239.62 μm). The measured depth of penetration of 12.4 mils and 9.4 mils equal 5.3% and 4.0%, respectively, material thickness loss of a 0.236 in. thick plate.

In Figure 69, the mechanical gouges made after UHPWJ in the abrasive blasted coupon were measured at 5.8 mils (148.19 un) and 14.5 mils (367.76 μm). The measured depth of penetration of 14.5 mils equals 6.1% material thickness loss of a 0.236 in. thick plate. The original abrasive blasted profile was 3.5 mils (88 μm).

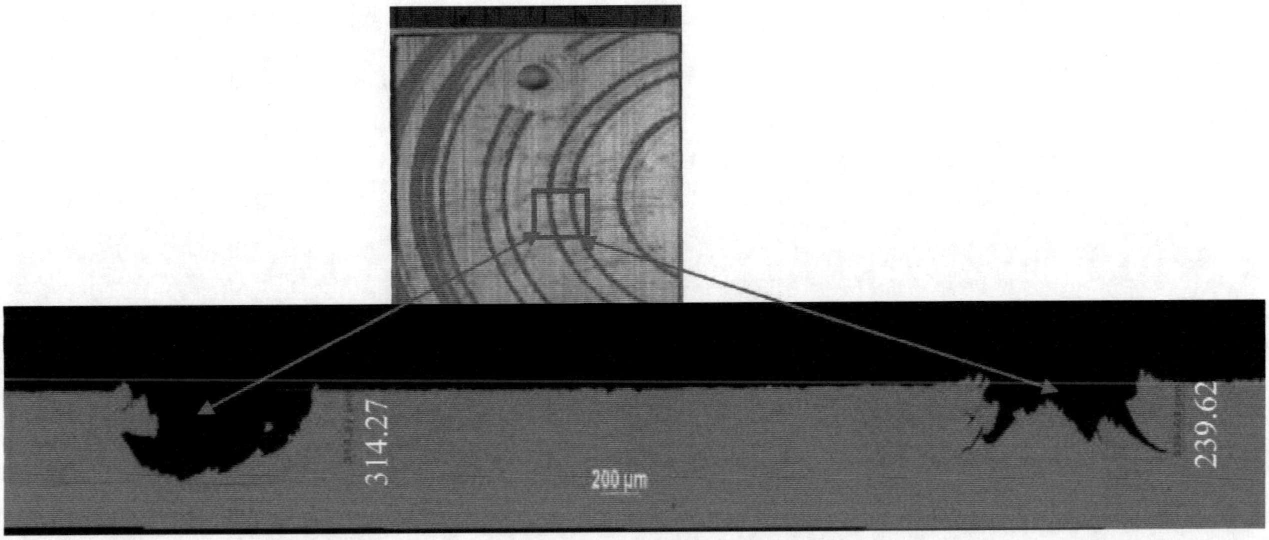

Figure 68. Metallographic image of a non-abrasive blasted mass loss coupon UHPWJ for 100 seconds in stationary position.

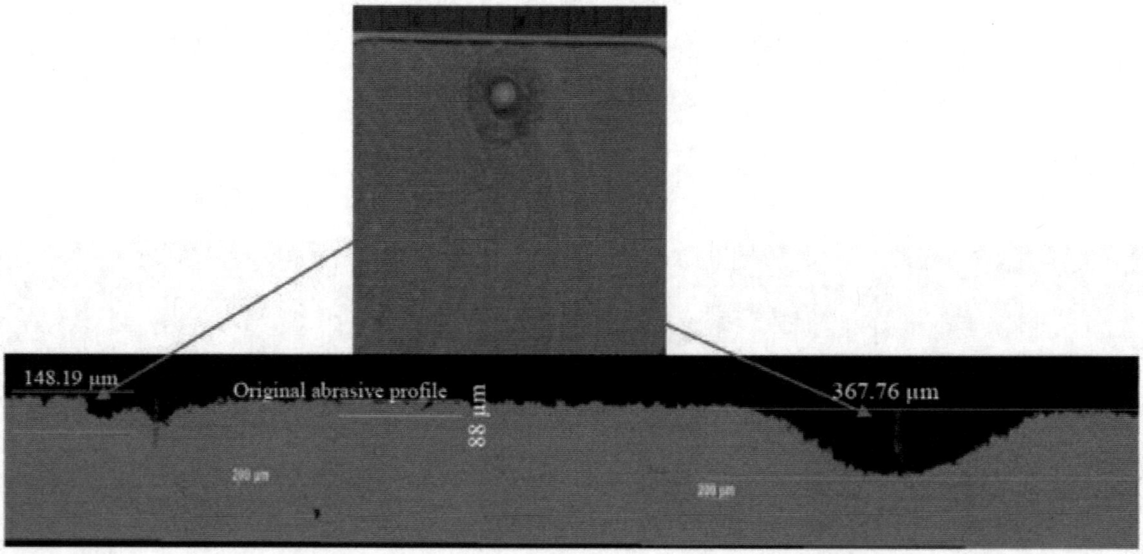

Figure 69. Metallographic image of an abrasive blasted coupon after UHPWJ for 100 seconds.

Representative metallographic images of a non-abrasive and an abrasive blasted Al 6082 coupon after UHPWJ for 120 seconds are shown in Figures 70 and 71. In Figure 70, the mechanical gouges made into the non-abrasive blasted coupon after UHPWJ were measured at 6.4 mils (163.01 µm), 6.2 mils (156.53 µm), and at 19.2 mils (488.73 µm). The measured depth of penetration of 19.2 mils equals 8.1% material thickness loss of a 0.236 in. thick plate.

In Figure 71, the mechanical gouge made into the abrasive blasted coupon after UHPWJ was measured at 19.5 mils (495.85 µm), which equals 8.3% material thickness loss of a 0.236 in. thick plate.

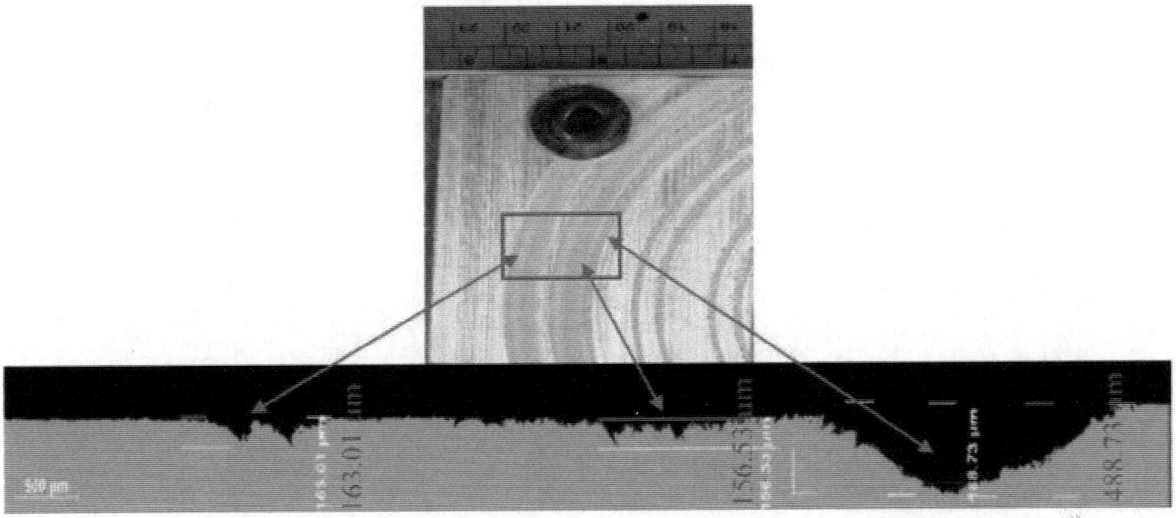

Figure 70. Metallographic image of a non-abrasive blasted Al 6082 coupon after UHPWJ for 120 seconds.

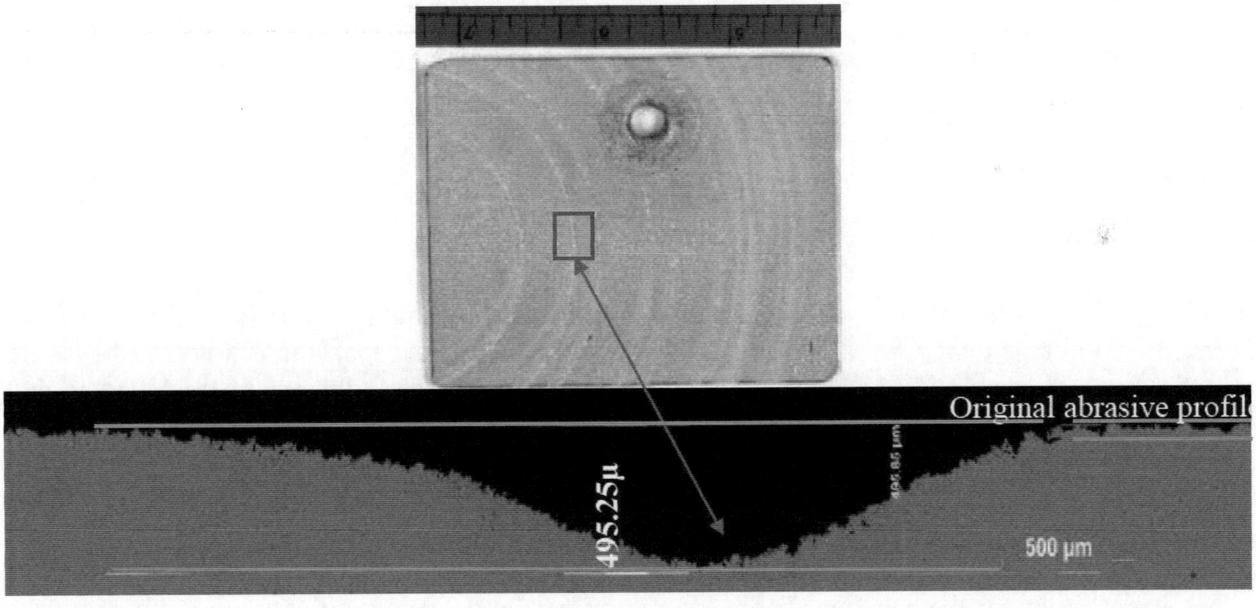

Figure 71. Metallographic image of an abrasive blasted Al 6082 coupon after UHPWJ for 120 seconds

Table 2. Maximum depth of penetration in relations to Al 6082 thickness loss per dwell time of UHPWJ exposure

Dwell Time of UHPWJ (seconds)	Maximum Depth of Penetration for Abrasively Blasted Coupons (Mils)	Maximum Depth of Penetration for Non-Abrasive Coupons (Mils)
120	19.5	19.2
100	14.5	12.4
80	8.8	8.9
60	12.8	4.2
40	4.7	9.3
20	4.8	8.6
15	3.1	3.2
10	N/A	1.3
5	N/A	1.3

Note: Table 3 in Reference 1

5.8 Adhesion testing

Representative photos of the adhesion testing via pull-off strength of coatings (ASTM 4541, Method E) and tape test (ASTM 3359, Method A). Due to the fact that this is a nonskid system, only the nonskid primer was used for adhesion testing. Results from adhesion testing showed that the primer coating on the steel substrate and aluminum alloys after six cycles of nonskid replacement retained good adhesion strength with the average values exceeded the minimum requirement of 400 psi by an order of magnitude of 10. The HY80-2 and Al 6061 plates that received twice as many UHPWJ passes than the HY80- 1 and Al 5456 plates during the trials showed no major differences in adhesion strength. Table 3 summarizes the adhesion pull-off strength test results of the baseline. Table 4 summarizes the adhesion pull-off strength test results after six cycles of nonskid application and removal.

Table 3. Summary of ASTM D4541, Method E (Adhesion Pull-off Strength) baseline results of both steel and aluminum panels.

Baseline			
Panel	**Avg Pressure at Failure (PSI)**	**Types of Failure: Cohesive/ Adhesive**	**Locations of Failure**
HY80-1	4150	Cohesive and Adhesive	Six dollies failed between glue and primer. Two failed between substrate & primer
HY80-2	3595	Cohesive and Adhesive	All dollies failed between glue and primer
Al 5456	2847	Cohesive and Adhesive	Five dollies failed between glue and primer. Three dollies failed between substrate & primer
Al6061	3309	Cohesive and Adhesive	All dollies failed between glue and primer.

Note: Table 4 in Reference 1

Table 4. Summary of the ASTM D4541, Method E (Adhesion Pull off Strength) results after six cycles of UHPWJ for both steel and aluminum panels.

After 6th Cycle of UHPWJ			
Panel	**Avg Pressure at Failure (PSI)**	**Types of Failure: Cohesive/ Adhesive**	**Locations of Failure**
HY80-1	4147	Cohesive and Adhesive	All dollies failed between glue and primer
HY80-2	3997	Cohesive and Adhesive	All dollies failed between glue and primer
Al 5456	2947	Cohesive and Adhesive	All dollies failed between glue and primer.
Al6061	2948	Cohesive and Adhesive	All dollies failed between glue and primer.

Note: Table 5 in Reference 1

5.9 SUMMARY NSWCCD-61-TR-2020/15

Table 5 summarizes the reduction rate per cycle of the peak and average surface profile of the steel and aluminum plates. Eleven cycles of UHPWJ would be required to remove over 3.0 mils of surface profile on aluminum flight decks, and 21 cycles for steel flight decks. For example, if the peak surface profile of an aluminum 6000 series flight deck was recorded at 4 mils, then after three nonskid replacements the surface profile would be below the minimum requirement of 3 mils called out in NSI 009-3.2

Table 5. Reduction Rates of the Surface Profile of Steel and Aluminum Plates

Material	Profile Reduction Rate (mils/cycle)	Net Peak Loss After 7 cycles (mils)	Net Average Loss After 7 cycles (mils)	Number of Cycles to Remove 3 mils of Profile
HY80-1	0.21	1.0	1.5	21
HY80-2 (14 passes of UHPWJ)	0.29	1.7	2.0	12
Al 5456	0.27	1.3	1.9	16
Al 6061 (14 passes of UHPWJ)	0.36	2.0	2.5	11

Note: Table 7 in Reference 1

Note: **Errata** HY80-2 Number of cycles is calculated at **12** rather than the published 21 in the Redacted Version of NSWCCD-61-TR-2020/15

Table 6 summarizes the reduction rate per cycle of the peak and average surface profile of the Al 6082 extruded structure. Seven cycles of UHPWJ would be required to remove over 2.0 mils of surface profile at the reduction rate recorded for Al 6082 flight decks.

Table 6 Reduction Rates of the Surface Profile of Al 6082

Material	Profile Reduction Rate (mils/ cycle)	Net Peak Loss After 3 cycles (mils)	Net Average Loss After 3 cycles (mils)	Number of Passes to Remove 3 mils Profile
Al 6082	.33	0.9	1.0	10

Note: Table 8 in Reference 1

Note: **Errata** Number of Passes to remove 3 mils Profile is calculated at **10** rather than the published 1 in the Redacted version of NSWCCD-61-TR-2020/15

Table 7 summarizes the reduction rate per cycle of material thickness of the steel and aluminum plates. The risk of removing steel and aluminum material over maximum thickness allowance via UHPWJ is very low. Deep gouges and circular grooves were revealed on the Al 5456 and Al 6061 plate surfaces during the waterjet process due to operator error as the ultra-high pressure water was impinging on the surface while the crawler unit was stationary. The measured depth of penetration of two of the deep gouges was approximately 40 mils.

Table 7 Reduction Rate of Material Thickness of Plates by UT Longitudinal Wave

Material	Material Thickness Reduction Rate (mils/cycle)	Net Maximum Material Thickness Loss After 7 cycles (mils)	Number of Cycles to Remove Over Maximum Thickness Allowance [10, 11]
HY80-1	0.21	1.5	912
HY80-2 (14 passes of UHPWJ)	0.29	2.0	912
Al 5456	0.36	2.5	620
Al 6061 (14 passes of UHPWJ)	0.43	3.0	517

Note: Table 9 in Reference 1

Metallography images of the HY80-1 steel plate samples revealed that the peaks became smoother after six cycles of nonskid replacement. Metallography images of the HY80-2 steel plate samples that received 12 passes of UHPWJ after six cycles of nonskid replacement revealed more of a rounded surface profile with areas of little profile. Metallography images of the Al 5456 plate samples revealed less frequent deep valleys and high peaks after six cycles of nonskid replacement, while the Al 6061 plate samples after 12 passes of UHPWJ revealed that periodic high peaks seen on the baseline profile images were generally not present. The Al 6061 surface profile appeared more ablated than the Al 5456 profile after 12 passes of UHPWJ.

Mass loss testing of Al 6082 resulted in a linear net increase in mass loss and material thickness loss as the dwell time of the crawler unit waterjetting at 40 ksi in a stationary position increased. While the original surface profiles for all plates decreased with increasing cycles of UHPWJ, the adhesion strength of the primer coating remained good with average pressures far above minimum requirement after six cycles of nonskid replacement.

6 PRODUCING A PROFILE

Metals have different threshold pressures, the pressure where the material is being removed. As early as 1995, patents by Van Kuiken and Byrnes were issued for pure Water Jets to generate a profile in malleable light metal comprising the steps of: roughening said surface by creating jets of water having pressures sufficiently high to clean and erode the surface to provide a pitted surface with undercuts, so that said surface is provided with a mechanical /adhesive bond for said coating…

6.1 VanKuiken Patent Malleable Light Metal

Van Kuiken and Byrnes were working between 35,000 and 55,000 psi. This patent finds application in engine block manufacturing and claims thermal spray coatings of aluminum, aluminum-bronze alloy, and low carbon alloy steel. Reference 15

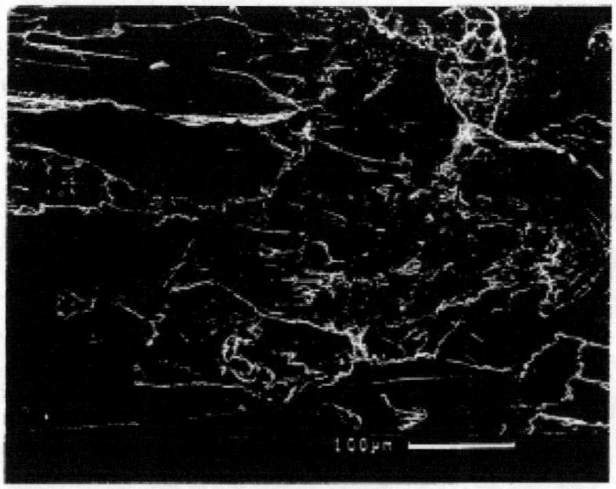

PRIOR ART

Fig. 3A

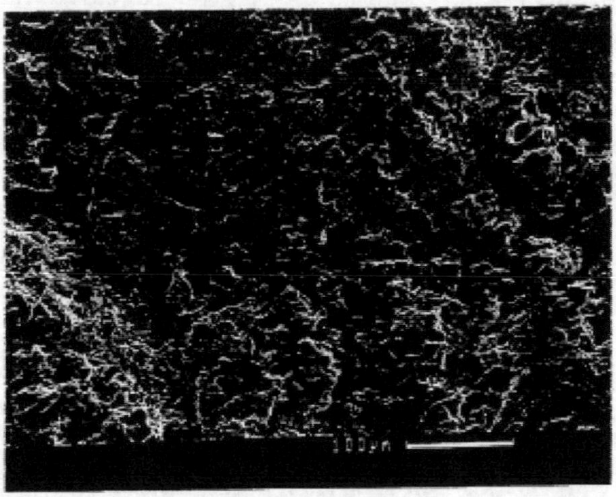

Fig. 3B

Figure 9- abrasive blast surface compared to WJ generated surface

Note: Figure 3 in Reference 15

The grit blast surface has high peaks to valley, and flat gouge surface; The thermal Spray coating adhesion is 3000 psi. The lower WJ surface has a 10-75 micron profile, much more peaks per unit area, the thermal spray coating adhesion is 6000 psi.

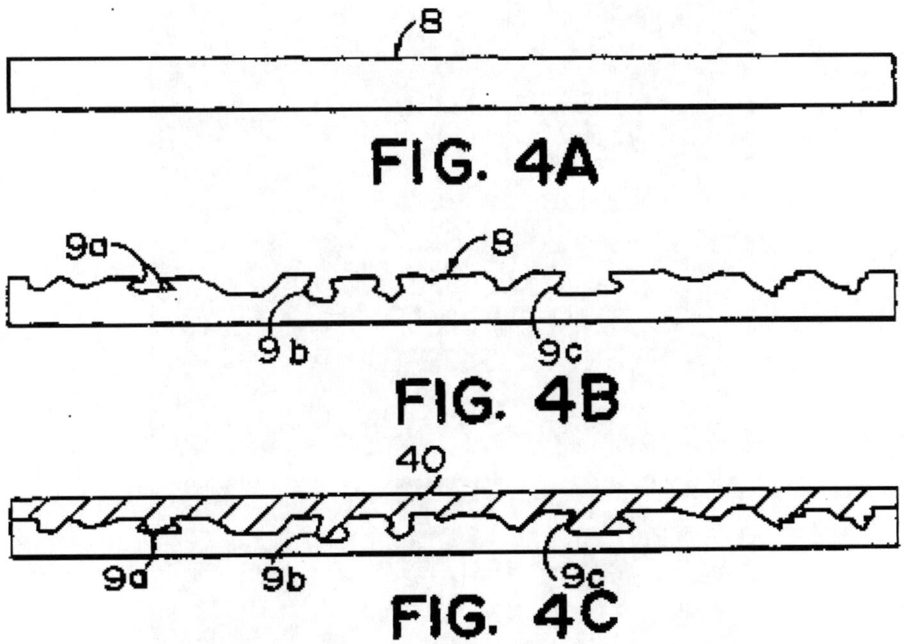

Figure 10. Depiction of cross section illustration the undercut for adhesion enhancement

Note: Figure 4 in reference 15

The adhesion of the thermal spray coating was enhanced by the undercutting of the water jet process.

6.2 Taylor Surface Roughening Inconel IN 718

The formation of profile and enhancement of adhesion was verified by Thomas Taylor in 1995. See Reference 16

In 1995, Taylor worked with a stream of 0.016 inch (0.4 mm) diameter orifice, standoff of 3 in (7.6 cm), with an erosion trace of about 0.06 in (1.5 mm). Mass loss is calculated as milligrams of mass lost per square centimeter of surface area. The substrate was Inconel IN-718. The measured erosion (mg/ cm^2) ranges 0.26 to 42.17; the calculated thickness loss in μm ranges from 0.1 to 51.2.

Two features of water jet erosion were noted:

- There is a minimum mass of water required to impinge before there is measurable erosion, that is, an incubation period.
- There is a threshold pressure required for measurable erosion.

The threshold pressure of IN-718 is about 28,500 psi (196 MPa) which is slightly under the 30,000 psi (204 MPa) pump.

Taylor's photographs are similar to VanKuiken's photographs.

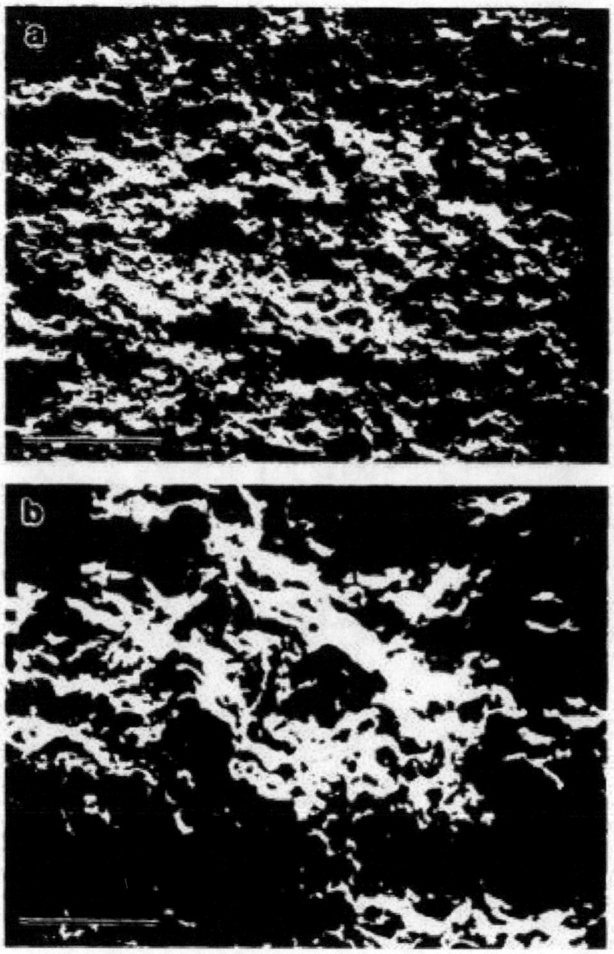

Figure 11 WaterJet profile of IN-718

The IN-718 eroded at 50,000 psi (345 MPa) The feature size is at least an order of magnitude finer in the waterjet surface.

The striking point is …the detail of the eroded surface increases with increasing magnification, suggest the waterjet erosion produces a fractal surface. … granular features of about 2 microns … rather micro-faceted.

Excellent bonding of a thermal spray overlay was obtained with this surface preparation having an absolutely clean interface.

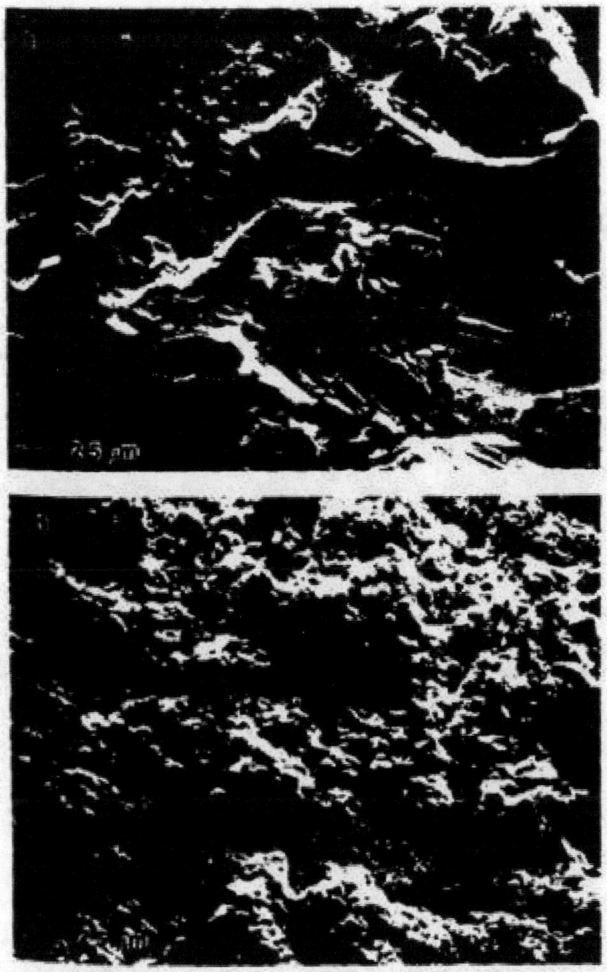

Figure 12 Alumina grit blasted surface.

The grit-blast surface would appear smoother as the magnification is increased, going from a long-range roughness pattern to smooth plateaus and facets, although there is micro-grooving due to the abrasion of the grit particle. The roughness of the grit blat surface isa about 5.3 μm, while the waterjet surface is about 6.0 μm, much the same in magnitude but substantially different in detail.

6.3 Vijay AL 6061-T6511 Aluminum Alloy

More recently, Vijay, Tieu, et.al. have reported on enhancing the adhesion strength of sprayed coatings. They start with an aluminum oxide abrasive blasted AL 6061-T6511aluminiumm alloy substrate. See Reference 17

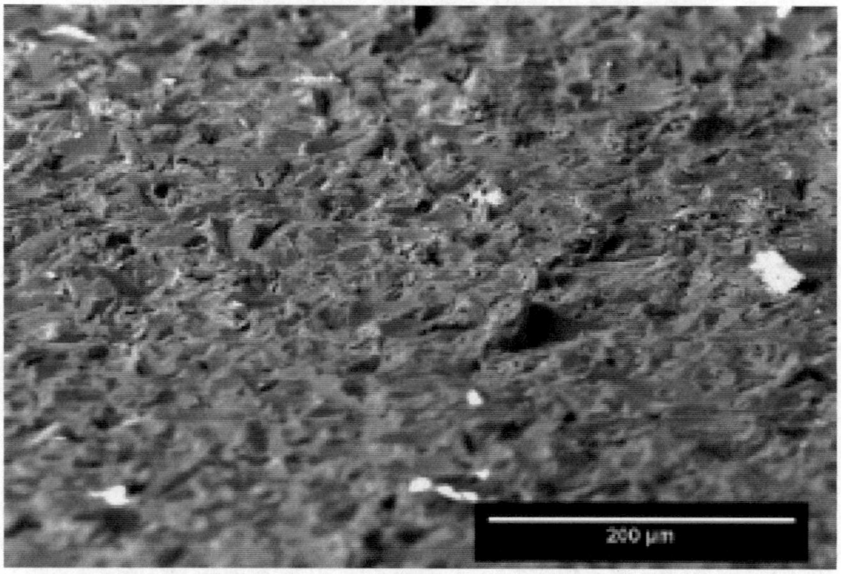

Figure 13: SEM image of grit blasted aluminum showing traces of alumina oxide grit embedment

Figure 1 in Reference 17

Figure 14: SEM Image (200x) showing PWJ roughened surface free from grit contamination.

Figure 2 in Reference 17

The surface topography is increased by random pits and crevices created by cavitation. The adhesion strength and surface roughness appear to have a direct correlation.

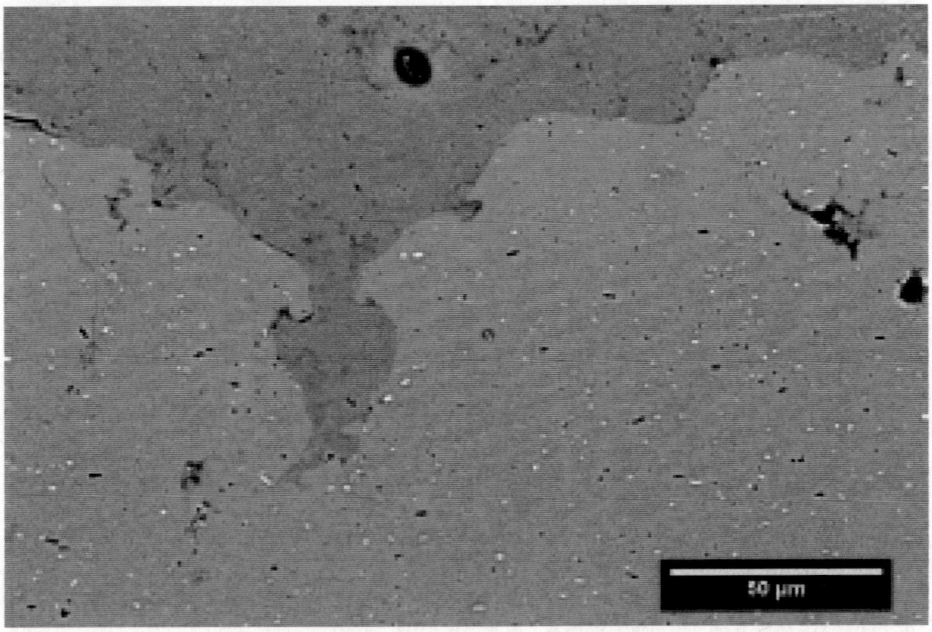

Figure 15: SEM image (500X) showing a cross sectional view of the coating and substrate interface. Note the interlocking effect offering strong mechanical anchoring.

Figure 6 in Reference 17

Vijay's conclusions are:

- Increased adhesion strength is a result of increased surface area and increased mechanical anchoring features.
- The surface roughness is beyond what can be expected from grit blasting.
- The additional benefit of not having embedded grit.

Vijay et.al. has also produced different profiles on biocompatible metals such as titanium to generate a progressive micro-and nano-topography with bioactive abilities for biomedical applications.

6.4 -Hashish Superplastic Formed Metals

Ramulu, Hashish and Chillman investigate the use of waterjets (>600 MPa) to qualify the material removal mechanism and form fine grain Ti-6Al-4V titanium alloy. The surface preparation for the formation of texture and removal was dependent on process dynamics. Reference 18

6.5 Hare Adhesion Factors

Hare published a series of articles on the fundamentals of paint for the Journal of Protective Coatings and Linings (JPCL) over several years. Clive Hare is my "GO TO" papers as Dr. Summers articles are my "GO TO" for waterjet fundamentals. The prior examples all verify the discussion of Clive Hare. Hare includes requirements for maximized adhesion in Reference 19.

For quality and longevity of paint performance, the substrate should have high surface energy, increased surface area, a pure surface without contamination, the presence of active sites, and porosity. See reference 19

Hare's summary list includes:

- Molecular bonding- mechanical bonding
- Expansion of surface area increases the number of potentially reactive sites for primary or secondary bonding.
- The thin, impermeable oxide layer that immediately reforms on newly bared iron surface is adherent and suitable as a substrate for good adhesion.

7 CONCLUSIONS

Water Jetting can produce a surface profile or texture on substrate.

The process must be automated so that the desired profile is achieved and not manifested as a deterioration.

The field use of 272 to340 MPa (40,000 to 50,000 psi) has exceeded the threshold pressure of steels as well as aluminum alloys and leads to material removal.

The undercutting of the substrate and the increase of the surface area are beneficial to increase the real surface area.

The exposed surface area for a water jet texture is greater per unit area than comparable abrasive blast surface area.

Thermal spray coatings work well on waterjet textures even if the profile is not as deep at abrasive blasted surfaces.

Practitioners (Jetters, Contractor) should be made aware that material is being removed when the robotic head is stationary and might need to change their work habits.

Authorities (Owners, Specifiers, Trainers, Coatings Experts, Coating Manufacturers) should consider that a conventional surface texture might not be required for performance. However, additional testing such as cathodic disbondment, or a lateral shear test rather than a perpendicular pull-off adhesion to determine if the macro peak-to-valley is required.

Do not let the head stop moving!

8 ACKNOWLEDGEMENTS

Thanks to the WJTA, HoldTight, Carolina Equipment and Supply, and the Advisory Council for sponsorship for this paper and a spot on the Back To Basics workshop. It is a real pleasure working with the WJTA staff and meeting global participants. Every individual brings something to the table and shares their point of view.

9 REFERENCES

1. Naval Surface Warfare Center, Carderock Division Technical Report, NSWCCD-61-TR-2020/15, July, 2020 The author is redacted, available by FOIA, redacted version available from Lydia Frenzel- with request by email; unredacted version is Distribution limited to U.S. Government Agencies and their contractors, request to Commander, NSWC, CD, Code 61, West Bethesda, MD

2. Material Characterization of the Modified Medium Pressure Water Paint Stripping Process: Composite Materials, Batelle Report, Richard Slife, et.al, WR-ALC-TIET, 1994, p 242 to 296

3. Roland Lever, "A Guide to Selecting Water Jet Equipment for Coating Installation Surface Preparation" NACE Infrastructure Conf., Baltimore, Fall, 1995

4. Water Jet Workshop, July, 1994, Bremerton WA, Sponsored by Naval Sea Systems Command Detachment, Planning and Engineering for Repairs and Alterations, Aircraft Carriers (PERA CV), organizers Bob Wheeler, Mike Gustavson. This workshop introduced the Ship Arms System™, The Amclean Hardware, concept of Flash Rust, Akzo Nobel photographs, Ameron-salts on surfaces.

5. Hutchens, Keen, Smith, Dillard, DeWeese, "Implementation of Environmentally Compliant Cleaning and Insulation Bonding for MNASA",
ntrs.nasa.gov/api/citations/19950025375/downloads/19950025375.pdf
Dillard, DeWeese, Hoppe, Vickers, Swenson, Hutchens," Evaluation of Pressurized Water Cleaning Systems for Hardware Refurbishment"
ntrs.nasa.gov/api/citations/19950025361/downloads/19950025361.pdf

6. NACE WJ-2/SSPC-SP WJ-2-2021, Waterjet Cleaning of Metals—Very Thorough Cleaning (WJ-2). Parallel text is found in NACE WJ-1/SSPC-SP WJ-1-2021, Waterjet Cleaning of Metals— Clean to Bare Substrate (WJ-1) and NACE WJ-3/SSPC-SP WJ-3-2021, Waterjet Cleaning of Metals— Thorough Cleaning (WJ-3

7. McGaulley, W, Shepperson, W., & Berry, Fred, "Comparison of Secondary Surface Preparation over Water Jetted Surface and Effect on Coatings Performance", SSPC 2001 Conference, p 9

8. Borkowski, "Influence of rotation Water Jet Kinematics on Effectiveness of Flat Surface Treatment," WJTA Conference, 2005, paper 5B-3

9. Kunaporn, Ramulu, Hashish, "Mathematical Modeling of Ultra High Pressure Waterjet Peening, paper 3-A, WJTA Conference, 2003

Kunaporn, Ramulu, Hashish, Hopkins, Ultra High Pressure Waterjet Peening Part 1: Surface Texture, paper 25, WJTA Conference 2001, 7075-T6 Aluminum alloy, P >= 150 MPa,

Kunaporn, Ramulu, Hashish, Hopkins, Ultra High Pressure Waterjet Peening Part II: High Cycle Fatigue Performance, paper 26, WJTA Conference 2001 Water Jet cleaning can influence the energy of the substrate surface and enhance fatigue strength.

Kunaporn, Ramulu, Hashish, "Mathematical Modeling of Ultra High Pressure Waterjet Peening, paper 3-A, WJTA Conference, 2003

10. Stein Bjoerndal, Ameron International, Paints and Coatings Group, Lierskogen, Norway, 1995, NACE CORROSION conference panel; private communication

11 Frenzel, L.M., Interim Report to Butterworth Co., L.M. Frenzel, R. De Angelis, J.B. Bates, "Evaluation of 20,000 PSI Water Jetting For Surface Preparation of Steel Prior to Coating or Recoating", April, 1983

L.M. Frenzel, "Application of High Pressure Water Jetting in Surface Preparation", National Paint and Coatings Association, Inc., Washington, D.C., Twenty-fifth Anniversary Marine and Offshore Coatings Conference, 1985

L.M. Frenzel, M. Ginn, G.V. Spires, Application of High Pressure Water Jetting in Corrosion Control "Surface Preparation: The State of the Art", B.R. Appleman and H.E. Hower, eds., Steel Structure Painting Council, Pittsburgh, PA, p. 164, 1985

L.M. Frenzel, Jonell Nixon, Pipeline Industry, "Corrosion and Pipe Protection; Prepare Steel Surfaces with High Pressure Water Blasting" Mar, p. 23, 1991

12 Dupuy, R, Howlett, J.J., "Ultra-High Pressure Water Jetting (UHP WJ): A Useful Tool for Deposit Removal and Surface Preparation", NACE Corrosion92, paper No. 253

13 R.K. Miller, G.J. Swenson, "Erosion of Steel Substrates When Exposed to Ultra-Pressure Waterjet Cleaning Systems, paper 52, WJTA Conference, 1999

14 Reference: Wright, Wolgamott, Zink, "Safe Waterjet Cleaning of Steel Process Lines", WJTA Conference, 2005, paper 2B-1

Wright, Wolgamott, Zink, "Parameters Affecting Surface Preparation", WJTA Conference, 2007, paper 2-A

Wright, "The Combination of Pressure and Flow Rate in the Expression of Relative Waterjet Impact", WJTA Conference, 2017,

15. Vankuiken, Byrnes, Kramer, US Patent 5,380,564 January 10, 1995,

16 Taylor, Thomas A. "Surface Roughening of metallic substrates by high pressure pure water" Surface and Coatings Technology, Vol 76-77, (1995) 22nd International Conference on Metallurgical Coatings and Thin Films, p 95-100

17 Vijay, Tieu, Yan, Daniels, Xu, Jordoin, Samson, MacDonald, Fernandez, Yandouzi, "Enhancing the Adhesion Strength of Cold gas Dynamic Sprayed Coatings by Preparing the Substrates with the High-Frequency Pulsed Waterjet", 2015, WJTA-IMCA Conference and Expo

Steeves, Variola, Vijay, Tieu, Yan, Xu, and Daniels, "In Vitro Activity of forced Pulsed Waterjet (FPWJ) Modified Titanium", 2017, WJTA-IMCA Conference and Expo

18 Chillman, Ramulu, Hashish," Investigation of Surface Preparation in Superplastic Formed Metals,", 2007, WJTA Conference, paper 3-D, SPF Titanium, 600 MPa

19 C. Hare, "Adhesion I", Journal of Protective Coatings and Linings (JPCL), 1996, May, p 77-86

C. Hare, "Adhesion: Part 2", JPCL, 1996, July, p. 79-95 JPCL articles are found at paintsquare.com

AUTHOR INDEX